Luxury Yachting

Tatiana Gladkikh · Hugues Séraphin ·
Victoria Gladkikh · Tan Vo-Thanh
Editors

Luxury Yachting

Perspectives on Tourism, Practice and Context

palgrave
macmillan

Editors
Tatiana Gladkikh
The University of the Highlands
and Islands
Perth, UK

Victoria Gladkikh
Independent Luxury Yacht Practitioner
Dunkeld, UK

Hugues Séraphin
Business School
University of Winchester
Winchester, UK

Tan Vo-Thanh
La Rochelle Business School
Excelia Group
La Rochelle, France

ISBN 978-3-030-86405-7 ISBN 978-3-030-86406-4 (eBook)
https://doi.org/10.1007/978-3-030-86406-4

This Palgrave Macmillan imprint is published by the registered company Springer Nature Switzerland AG
The registered company address is: Gewerbestrasse 11, 6330 Cham, Switzerland

Acknowledgments

As the editors of this book, we should like to thank all the authors who have contributed their chapters to this monograph. Your academic insight and professional knowledge have made this publication a unique collection of perspectives which you hold on luxury yachting. Your knowledge, hard work and cooperation are very much appreciated.

Our special Thank You goes to Palgrave Macmillan for responding with enthusiasm to our idea of the book and for the detailed guidance, support and cooperation at every stage of the process.

As a lead editor of the book, I would like to say a big thank you to our editorial team: to Victoria for her luxury yachting expertise and access to the industry, and to Hugues and Tan, my trusted colleagues, for their ideas and support with reviewing the chapters. Working with a professional and responsive team has made this project very efficient and enjoyable. Thank you for another book produced together!

Of course, we should like to acknowledge all those luxury yachting professionals who have given their invaluable time and knowledge of the industry to our authors and shared their rich accounts of life at sea for the benefit of our better understanding of the world of luxury yachting.

Contents

Part IV Conclusion

Notes on Contributors

Lyn Batchelor is Director of Studies and Senior Lecturer at the University of Aberdeen's Campus in Qatar. She has worked previously in universities in Scotland, England and Australia. She teaches entrepreneurship and is interested in how innovation can inspire new models for existing businesses to grow and thrive. Her Ph.D. was framed by Critical Management Studies and dealt with how people in a luxury service industry used language as part of their narratives of self to impression manage and position themselves in a crowded market. She has served on many committees in higher education and the wider community, and worked with start-ups and female entrepreneurs on various European projects.

Emiel L. Eijdenberg is a Senior Lecturer at James Cook University, Singapore. Dr. Eijdenberg's research interests are about entrepreneurs(hip) and elite tourism, especially in the tropics. Dr. Eijdenberg's work has been presented at international management conferences such as ANZAM, AoM, ESU, EURAM, ICSB and RENT.

Tatiana Gladkikh is a Sector Manager—Business, Management and Computing—at the University of the Highlands and Islands (UHI), Perth College. As an academic and consultant, she has worked in the UK, Russia, Europe and the Middle East. Tatiana teaches at all academic levels and researches leadership, globalization and cross-cultural issues in the global business environment. Her areas of consultancy are cross-cultural business communication and transnational education. Tatiana holds a Ph.D. from the University of Birmingham. Among her recent publications are *The International Business Environment and National Identity* (Routledge, 2018) and *Overtourism—Causes, Implications and Solutions* (Palgrave Macmillan, 2020). Dr. Gladkikh is a Senior Fellow of the Higher Education Academy (SFHEA) and a Member of the Chartered Institute of Linguists (MCIL).

Victoria Gladkikh is a luxury yachting practitioner. Having worked in the industry since 2016, she has acquired over 20,000 nautical miles travelling on luxury vessels of various sizes around Europe, North America, Canada, the Caribbean and Central America. She has served a variety of affluent owners and guests from various cultural and professional backgrounds. She has experienced two transatlantic crossings, the Panama Canal crossing and Monaco Yacht Show. Victoria is a qualified Yachtmaster. A recipient of the Slater Scholarship award, she is currently studying for an Officer of the Watch qualification. Victoria has two significant media appearances: her professional interviews and photographs were published in leading luxury yacht magazines *On Board* and *Boat International*. She is an active member of the female yachting organization *She of the Sea*.

Janti Gunawan is a Senior Lecturer at the Department of Business Management, Institut Teknologi Sepuluh Nopember (ITS), Indonesia, with research interests in local economic development, tourism and international business. She has both engineering and commerce degrees from Indonesia, Australia and New Zealand.

Andrew Jones holds a senior academic post at the University of Malta and has also held positions at The University of Wales, International Faculty Cardiff UK, Swansea Business School, UK and the University of

Brunei. Andrew has professional experience in planning, environmental management and tourism at both international, regional and local levels and has been an enthusiastic contributor, for over thirty years, to the academic tourism community within Malta, the United Kingdom and internationally. Andrew first qualified in Urban and Regional Planning, but joined higher education in the late 1980s where he focused his research on tourism, sustainability and environmental planning. In his most recent work, Andrew has focused part of his teaching and research specializing in two distinct areas: (i) sustainable tourism—specializing in community and niche market development, coastal tourism and climate change and (ii) the development of transnational education (TNE). In this respect, he has had the opportunity to work within academic sectors internationally in Europe, Central, South and Southeast Asia. These include assignments in Japan, China, India, Hong Kong, Singapore, Uzbekistan, Malaysia, Japan, Germany Greece. Italy, Bulgaria, Romania and the United Kingdom over the last twenty years.

Muhamad Azlian Kautsar is a Bachelor Student of Marine Engineering Department, ITS, Indonesia, with working experience as an oceanographic surveyor.

Aise Kim is a Senior Lecturer in the School of Management at the University of South Australia, Adelaide. She completed her Ph.D. research on sustainable tourism management at the University of Surrey, UK. Her research interests include environmental behaviour, interpretation, tourism marketing and sustainable tourism.

Thean Labuschagne is a luxury yacht chef who has been working in the industry from 2017. Before joining the yachting world, he has gained significant experience through working in some of the top ten restaurants in South Africa (2011–2016). Thean studied at Culinary Alfresco International Academy of Food and Wine, with Chef Annette van der Water as its Head Chef and Principal. Through his training Thean learnt the importance of conscious cooking from an outset of his career. Since then, he has been striving towards being conscious of his environment with a goal to become known as an ambassador for sustainability in the yachting industry.

Caroline Navarro worked as a chef on superyachts for 10 years. She completed a Master in Tourism at the University of Malta with a thesis entitled 'Yacht racing as a niche tourism product: Case study the Rolex Middle Sea Race 2012'. In 2013, Caroline presented her master paper at the ATLAS Conference and got it published by *Emerald—International Journal of Events & Festival Management—Events & the Blue Economy*. Caroline worked for the Camper & Nicholson Marinas in Malta, the Grand Harbour Marina as a sales & marketing manager for 5 years. In 2018, Caroline chose to advise on specialized yachting projects concerning a new crew concept and a marina & regatta in the Balearic Islands. Since she has presented at the Quaynote Conference 'Opportunities in Superyachts' in Palma de Mallorca and Malta.

Lauren Ryburn has been employed as a stewardess within the yachting industry for over seven years and has worked on board various sized vessels ranging from 45 to 98m, travelling to many different locations around the globe. In her experience she has held a range of positions: 2nd Stewardess, Stewardess, Stewardess/Spa therapist, Stewardess/Masseuse. In 2018 she was a pioneer in establishing a yacht-crew targeted blog and Instagram account called 'The Green Stew' which promotes sustainability within the Yachting Industry. Lauren received the Acrew Environmental Crew Member of the Year award 2020 for her environmental efforts within the industry. Lauren's written work and other content include:

Instagram page (@the.green.stew): https://www.instagram.com/the.green.stew/?hl=en Clear Ocean Pact article: https://clearoceanpact.org/spotlight-on-the-green-stew/.

Seastainable Yachting article: https://www.seastainableyachting.com/post/q-a-withthe-green-stew. The Green Stew website and blog: https://www.thegreenstew.com/.

Hugues Séraphin is a Senior Lecturer in Event and Tourism Management Studies. He is also an associate researcher at La Rochelle Business School (France). Hugues Séraphin holds a Ph.D. from the Université de *Perpignan Via Domitia* (France) and joined the University of Winchester Business School in 2012. He was the Programme Leader of the Event Management programme between 2015 and 2018. Dr. Hugues Séraphin has expertise and interests in tourism development and management in postcolonial, post-conflict and post-disaster destinations. He has recently published in *International Journal of Culture, Tourism, and Hospitality Research*; *Current Issues in Tourism*; *Journal of Policy Research in Tourism, Leisure and Events*; *Journal of Business Research*; *Worldwide Hospitality and Tourism Themes*; *Tourism Analysis*; and *Journal of Destination Marketing & Management*. https://www.winchester.ac.uk/about-us/leadership-and-governance/staff-directory/staff-profiles/seraphin.php.

K Thirumaran is a Senior Lecturer at James Cook University, Singapore. His research interests focus on heritage, culture, luxury and transformational tourism. Thirumaran coined the term 'affinity tourism' which refers to the propensity of guests to partake in the 'familiar' and 'similar' cultural experiences to those of their hosts.

Katrien van der Merwe hails from South Africa. Prior to her yachting career, she earned a B.A. degree in Psychology and an Honours degree in Business. She has been in yachting for over eight years and have worked on various vessels ranging in size, from 40m to over 100m. During her career she has visited over 42 countries. She is a senior crew member who has worked with various captains and crew from different nationalities and cultural backgrounds.

Tan Vo-Thanh is an Associate Professor in Tourism & Marketing at La Rochelle Business School, Excelia Group, France. His area of specialization and research is Destination Management, ICT, Wine Tourism, and Green Human Resource Management.

Amien Widodo is an Associate Professor at the Department of Geophysics Engineering, ITS, Indonesia. As a geologist, his research interests are in the area of geo-hazard and geo-tourism.

Russell Williams is a Senior Lecturer at the University of Aberdeen and currently serves as Dean for the Qatar campus, a project he has been engaged with over the last 5 years. An experienced academic, trainer and researcher having worked at a number of UK Universities, Russell's teaching interests range across Business Strategy, Leadership, Strategic Marketing and Digital Business. Russell has published widely across a range of journals including: *Marketing Intelligence and Planning*, *Management Decision*, *Managerial Auditing Journal*, *Organization Development*, *Management Decision*, *Journal of Electronic Commerce Research*, *International Journal of Technology & Human Interaction*, etc. In a previous life, Russell was involved in two Internet start-ups at the turn of the millennium. He has also served as a business mentor for start-ups.

Caroline Wong is the Associate Dean of Learning and Teaching and Senior Lecturer at James Cook University, Singapore. Dr. Wong's research in knowledge management takes on a multidisciplinary approach that extends into knowledge-based cities, smart cities, creative cities and creative industries with a special focus on Singapore.

Karen Worrall is a content writer, copywriter and international sales and marketing manager. She has worked on luxury cruise ships since 2005 in entertainment, and later as sales and marketing manager since 2019. Her experience includes six leading luxury cruise lines and a luxury yacht company. Karen has been writing content and copywriting since 2017. She has worked in-house for National Geographic Traveller (UK) as an editorial intern and created her own cruise travel site Cruiseshipkaren.com. Prior to her employment on ships she worked in France, Spain and Italy, where she learned the languages of these countries. Karen holds a B.A. in Applied Music and is currently studying for an MBA Executive at the University of the Highlands and Islands.

List of Figures

List of Tables

1

Introduction: Luxury Yachting—a Growing but Largely Unknown Industry

Tatiana Gladkikh, Hugues Séraphin, Victoria Gladkikh, and Tan Vo-Thanh

This book offers its readers an insight into the luxury yacht industry as a provider and facilitator of a luxury yacht experience. Luxury yachting is an exclusive area of tourism and practice which operates in a relatively small and niche environment. Often concealed from the broad

T. Gladkikh (✉)
The University of the Highlands and Islands, Perth, UK
e-mail: Tatiana.Gladkikh.Perth@uhi.ac.uk

H. Séraphin
Business School, University of Winchester, Winchester, UK
e-mail: hugues.seraphin@winchester.ac.uk

V. Gladkikh
Independent Luxury Yacht Practitioner, Dunkeld, UK

T. Vo-Thanh
La Rochelle Business School, Excelia Group, La Rochelle, France
e-mail: vothanht@excelia-group.com

T. Gladkikh et al. (eds.), *Luxury Yachting*,
https://doi.org/10.1007/978-3-030-86406-4_1

public attention, it attracts a very special type of clientele with high-end income and high social status (Katsioloudes & Hadjidakis, 2007). However, despite being small and niche, it operates on a global scale and the number of newly built yachts is rapidly growing. According to Statista, the value of the luxury yacht and private jet market worldwide has grown from 21 billion Euros in 2018 to 24 billion Euros in 2019 (www.statista.com, 2020). The Times (2020) has recently reported with reference to Boat, the private yacht 'bible', that the private yacht sales 'are tracking up 10%' compared with the same quarter last year. Analysts expect the sales to break the record over the next few months. The surge in sales has been accelerated by the desire to escape COVID-19 on what seems to be the safest place on the planet at the time of pandemic (The Times, 2020).

This book introduces a new form of special interest tourism (SIT), namely Luxury Yachting Tourism or Superyacht Tourism. It contributes to a wider range of SIT which also includes gastro tourism, atomic tourism, begpacking, booze tourism, Christina tourism and Halal tourism. The very detailed list of SIT does not include Luxury Yachting Tourism. How does this form of tourism fit within the existing definition of SIT? Special Interest Tourism (SIT) can be observed 'when the traveler's motivation and decision-making are primarily determined by a particular special interest with a focus either on activity/ies and/or destinations and settings' (Trauer, 2006: 186). SIT appears to accommodate the varied and specialized needs and tastes of tourists and is to be opposed to mass consumption and non-commercialized individual travel (Park, 2014; Trauer, 2006). This form of tourism emerged in the 1980s (Trauer, 2006) and was stimulated by a need for cultural and environmental holidays. SIT contributes to enhancing the image of a destination, enriching tourists' experiences and is profitable to a wider range of providers (Jin & Sparks, 2017). Other terms used alongside SIT are: alternative, sustainable, appropriate, new, responsible, eco, niche and ego tourism (Jin & Sparks, 2017; Trauer, 2006).

Currently the literature of luxury yachting is highly limited. The book search on www.amazon.co.uk only returns results on vacation planners, cruise diaries and generally yachting but not on luxury yachting specifically. Numerous publications can be found on luxury products, luxury fashion, luxury end of various industries but almost nothing appears on superyachts and luxury yacht tourism. *Sea the World Differently* by Todd Beechey (2017) and *The Law of Yachts and Yachting by* Coles and Lorenzon (2018) are the closest publications to our book on superyachts but they focus on the experience of holidaying on luxury yachts discussed in *Sea the World Differently* and the legal perspective adopted by *The Law of Yachts and Yachting* offers a very different angle and therefore these books find themselves in a very different category.

Methodology

This book aims at filling the gap in luxury yachting tourism literature and offers a unique approach of combining academic investigations of the industry with practitioners' perspectives on daily realities in the world of luxury yachting. Part I adopts more traditional academic research methodologies underpinning their investigations by already existing theoretical frameworks. In Part II an experiential evidence-based approach builds on the professional experience and judgment of four luxury yachting practitioners (Barends et al., 2014) where primary evidence is generated through our contributors' first-hand experiences. Additionally, it utilizes subjective personal introspection (SPI) (Holbrook, 1995) into the lives of crew who are currently working in this sector. The chapters in Part III are informed by a synthesized methodological approach of the two previous chapters. The methodology of this section of the book utilizes the codes of the Delphi method, without a *stricto* sensu adoption of its protocol. The Delphi method is characterized by the collection and interpretation of experts with a wide range of experience and a diversity of opinions in responses to the questions raised (Hammond & Wellington, 2013). In this book, the Delphi technique brings the real-life insight into the industry through the lens of those who have experienced it from the service provider perspective. Thus, in addition to the value brought from academic and practitioner perspectives on luxury yachting, this book also adds to the methodological developments in researching social reality in the post-modern world.

From a philosophical perspective widely accepted in business research, the book has adopted a Critical Research paradigm (Myers & Klein, 2011) which has been found very helpful in supporting the introspection into how existing professional/business practices influence people and communities. This method of knowledge creation is linked to and supported by the ethnographic approach of Lifecourse of the researcher, which brings together personal and professional experiences in connecting the researcher and the researched (Fois, 2017). The Lifecourse paradigm allows researchers to immerse themselves in insightful investigations of life journeys, critical points, influences and trajectories (Elder et al., 2003). This has proved very helpful in uncovering the reality of life of luxury yacht crew through the authors' exploration of luxury yachting experiences.

Utilizing one's personal insight in knowledge generation is argued by De Ascaniis and Grecco-Marasso (2011) to be valuable and trustworthy, as personal experience and consequent reflection enables the reader to engage with the tale of life experience. This can be related to travel writing as it depicts the tale of a journey and offers witness value (Séraphin & Mansfield, 2017). Overall, bringing practitioner experience and expertise gained from working in the industry is a useful approach in the book. The invaluable practitioner perspective sheds light into the world of luxury yachting and helps us achieve the goal of shaping a better understanding of the industry through practice-based theory building. As a helpful way of knowledge generation it enables bridging the gap between rigour and relevance of academic–practitioner relationships (Bartunek & Rynes, 2014).

The Structure of the Book

The book consists of an Introduction (this Chapter), three Parts (Chapters 2–11) and a Conclusion (Part IV, Chapter 12). This chapter, Introduction, offers the rationale for writing this book, discusses its methodological approach for researching the issues raised and provides a brief overview of each chapter. Conclusion, Chapter 12, synthesizes the findings of the book, highlights its main arguments and raises further questions stimulated by the discussions in this publication.

Part I focuses on luxury yachting from a tourism perspective and offers academic contributions evaluating the industry in the context of COVID, from a perspective of value proposition and co-creation of customer experiences, and in contrast to luxury cruising. Chapter Two, a conceptual study based on secondary research, proposes that the breakout of COVID-19 has offered the tourism industry an opportunity to review its current practice. It investigates the benefits and limitations of this sector, and its level of resilience. Chapter 3 builds on a reflexive account of a luxury yacht charter experience and offers a discussion on the importance of co-creation of experiences and the role of authenticity in experiences. Chapter 4 analyses the differences between luxury yachting and luxury cruising. Relevant theoretical knowledge on luxury, with a focus on signalling wealth, privacy and exclusivity, underpins the discussion and helps to untangle the similarities and differences of sea holidays, as well as customer preferences of choosing one over another.

Part II of the book further develops its interpretivist nature by providing an insight into the industry from the practitioner perspective. It includes narratives from luxury yacht professionals who have experienced/are experiencing the daily reality of working on luxury yachts and the industry more broadly. Their first-hand information enlightens the reader through their discussions of luxury yacht practices, successes and challenges in a complex intercultural context of the yacht crew. Chapter 5 reflects on the portrayal of the luxury yachting industry in the media and contrasts this with first-hand experiences from crew members actively working on yachts. It discusses the main responsibilities of the Deck, Interior and Engineering departments from a practitioner's perspective, and addresses a range of topics, including transportation to and from the vessel, communications between guests and crew, serving guest requirements including tourist experiences, entertainment and water sports.

Chapter 6 considers environmental practices of day-to-day operations and proposes an evaluation of the development of the move towards sustainability in the industry and discusses the ways of furthering green experiences for crew and guests. Chapter 7 discusses the differences between the work of land-based and luxury yacht-based chefs and the challenges they face respectively in order to work sustainably. It emphasizes the peculiarities of day-to-day luxury yacht chefs' operations and demonstrates the culinary side of luxury yachting in order to better understand the yachting industry as a whole. In Chapter 8 the author reflects on her own professional experiences and highlights challenges and advantages typically encountered by the crew which evoke a range of attitudes and perceptions of the industry among yachties, often experienced on a scale of contrasting and contradicting emotions, localities and memories. The author also acknowledges the impact of the industry on local communities in yachting destinations. We would like to highlight again the uniqueness of Part II of the book which lies in its real-life accounts written by practitioners.

Part III stimulates the discussions of the context of luxury yachting and offers discussions of the industry's practices and challenges from the point of view of destinations and engagement with local economies and communities. It discusses the regional treatment of localized tourist hubs. Chapter 9 investigates the marine tourism conditions in Indonesia before the COVID-19 pandemic and identifies the potential and challenges for developing sustainable geo-marine tourism as a tool for recovering from the COVID-19 pandemic crisis, while building the path to serve luxury yachters and suggests key strategies to maximize socio-economic benefits of sustainable marine tourism through collaboration among stakeholders and adopting a participatory management approach. Chapter 10 examines the significance and potential of yachting tourism in Singapore using the 'policy mobility' lens. The findings indicate the need for a more pronounced policy promoting luxury yachting tourism and proposes approaches to invigorate the luxury yachting segment in Southeast Asia to a level similar to the Caribbean and European Mediterranean Seas. In response, Chapter 11 discusses the Rolex Middle Sea sailing event as a specific luxury tourism example in Malta. It evaluates current and future contributions that such an event can bring to the local tourist economy and evaluates broader lessons that such events could offer for other destinations.

Chapter 12, Conclusion, highlights the global outlook of the book afforded through the academic and practitioner insights obtained in different parts of the world. It reflects on the methodological uniqueness of this publication and stresses its contribution to expanding our knowledge of the luxury yachting industry through the rigour of academic considerations and invaluable practitioner perspectives. The mentioning of a specific mindset, characteristic to luxury yachting and raised in various chapters, allows us to pose a question whether the industry with its global span of operations and truly intercultural and transnational character can be seen as a platform for addressing global challenges by influencing the development of a global mindset among its stakeholders.

References

Barends, E., Rousseau, D. M., & Briner, R. B. (2014). *Evidence-based management: The basic principles.* Centre for Evidence-Based Management.

Bartunek, J. M., & Rynes, S. L. (2014). Academics and practitioners are alike and unlike: The paradoxes of academic-practitioner relationships. *Journal of Management, 40*(5), 1181–1201.

Beechey, T. (2017). *Sea the world differently.* OceanScape Yachts.

Coles, R., & Lorenzon, F. (2018). *The law of yachts & yachting (Maritime and Transport Law Library).* Routledge.

De Ascaniis, S., & Grecco-Morasso, S. (2011). When tourists give their reasons on the web: The argumentative significance of tourism related UGC. In R. Law, M. Fuchs M., & F. Ricci (Eds.) *Information and communication technologies in tourism 2011* (pp. 125–137). Springer.

Elder, G., Johnson, M., & R Crosnoe, R. (2003). The emergence and development of life course theory. In T. Mortimer & M. Shanahan (Eds.), *Handbook of the life course.* Kluwer Academic Publishers, Hingham. Cited in Fois, F. (2017). Understanding ethnography through a life course framework: A research journey into alternative spiritual spaces. *Area, 49*(4), 421–428.

Fois, F. (2017). Understanding ethnography through a life course framework: A research journey into alternative spiritual spaces. *Area, 49*(4), 421–428.

Hammond, M., & Wellington, J. (2013). *Research methods: The key concepts.* Routledge.

Holbrook, M. B. (1995). *Consumer research: Introspective essays on the study of consumption.* Sage.

Jin, X., & Sparks, B. (2017). Barriers to offering special interest tour products to the Chinese outbound group market. *Tourism Management, 59*, 205–215.

Katsioloudes, M., & Hadjidakis, S. (2007). *International business: A global perspective*. Elsevier.

Myers, M. D., & Klein, H. K. (2011). A set of principles for conducting critical research in information systems. *MIS Quarterly, 35*(1), 17–36. Cited in Venable, J. (2011). Incorporating design science research and critical research into an introductory business research methods course. *Electronic Journal of Business Research Methods, 9*(2), 119–129.

Park, Y. P. (2014). *Heritage tourism*. Routledge.

Séraphin, H., & Mansfield, C. (2017). A Comparative study of tourism and hospitality education in the United Kingdom and France. *Journal of Applied Ceremonial and Communication in Management, 2*(2), 55–95.

The Times. (2020). Coronavirus: Private jet travel soars as wealthy flee high-risk areas. Retrieved 28 June, 2020, from https://www.thetimes.co.uk/article/coronavirus-private-jet-travel-soars-as-wealthy-flee-high-risk-areas-32xrk89xd

Trauer, B. (2006). Conceptualizing special interest tourism. Frameworks for analysis. *Tourism Management, 27,* 183–200.

Statista.com. (2020). Global value of the luxury yacht and private jet market 2019. Retrieved 28 June, 2020, from https://www.statista.com/statistics/675373/global-value-of-the-luxury-yacht-and-private-jet-market/

Part I

Perspectives on Tourism

2

Luxury Yachting in the Global Context of COVID-19

Hugues Séraphin

Introduction

Since the breakout of COVID-19, a significant body of research has been published on the impacts of the pandemic on the tourism industry and related sector (Séraphin, 2021; Sharma & Nicolau, 2020; Sigala, 2020; Tiwari et al., 2020). There are also many ongoing other projects, such as call for papers on the impacts of COVID-19 on tourism education (JHLSTE, 2020 [Online]). Having said that, so far none of the existing publications and planned projects focus on the luxury branch of tourism, let alone on luxury yachting. By discussing this sector of the tourism industry this study fills the gap in literature, not only in the field of tourism but also in the area of luxury products and services. In order to bridge the gap between the pandemic and yachting, this chapter firstly provides an overview of the impacts of COVID-19 on

H. Séraphin (✉)
Business School, University of Winchester, Winchester, UK
e-mail: hugues.seraphin@winchester.ac.uk

T. Gladkikh et al. (eds.), *Luxury Yachting*,
https://doi.org/10.1007/978-3-030-86406-4_2

the tourism industry. It then presents luxury yachting as a sector. Finally, the study discusses this sector in relation to the pandemic. As the impact of COVID-19 is still an unfolding topic (Jamal & Budke, 2020), online media sources such as yachting websites, online newspapers and YouTube are used to collect data for this study. This chapter concludes that luxury yachting is more than a form of tourism, it is also a driver for tourism development.

COVID-19 and Tourism

Tourism and Cognate Sectors

In March 2020, the world was hit by COVID-19, which subsequently led to thousands of deaths (Yang et al., 2020). This pandemic is considered as 'the first real global health emergency experienced by many members of today's society' (Jamal & Budke, 2020: 2). As a result, national and international travel was forbidden, and planned events were cancelled. These were among the first governments' actions worldwide (Jamal & Budke, 2020; Yang et al., 2020). Existing literature on COVID-19 and tourism could be split into two main categories: those focusing on the negative impacts of the virus on the tourism industry and related industries, and those perceiving the pandemic as an opportunity to review the development and management of the tourism industry (Séraphin, 2020). Amongst the obvious and natural negative impacts of COVID-19 on the industry is the economic impact on all sectors (e.g. travel agency, tour operators, transporters, accommodation, catering, events, cruises, etc.).

As for the positive potential outcomes, the breakout of the virus seems to have operated as a *desengaño* (eye opener) on many aspects of the operation of the industry, on training provision for future leaders and the sustainability strategies currently in place (Séraphin, 2020). In a nutshell, existing literature simultaneously looks at the past (before COVID-19) and the present, and to some extent the future (COVID-19 time, and post-COVID-19). This approach which tends to look at opposite directions at the same time is referred to as Janusian thinking, or ambidexterity (Mihalache & Mihalache, 2016). In tourism, ambidextrous management which is about adopting a management approach which puts together contradicting strategies, such as for instance the development of resorts (enclaves) which are open to visitors, with the overall purpose (among other things), to bridge the gap between locals and visitors (Séraphin & Butcher, 2018; Séraphin & Yallop, 2019). This management strategy has been associated with sustainability, performance improvement and so on. (Vo-Thanh et al., 2020). The events sector, which is directly correlated with the tourism industry, has also been hit severely by the pandemic (Séraphin, 2020).

Events Management Specific

When it comes to the event industry, the pandemic has mainly impacted the happenings of rites and rituals (Séraphin, 2020). The impacts were rather limited when it comes to rites of consumption and rites of exchange. They were particularly important for rituals of drama, rites of competition and de-valorization rites. However, the pandemic did not have any impacts on rites of purifications, rites of passage and rites of conspicuous display (Séraphin, 2020). Even if the impacts of the pandemic have been different according to the sector and according to the type of rites or rituals performed, the fact is that social capital among participants, service providers and such like could be considered as a common thread (Séraphin, 2020). It is worth highlighting the fact social capital is to be associated with such notions as inclusion in networks, reciprocity and trust and quality interaction (Getz & Page, 2020; Jamal & Budke, 2020; Séraphin, 2020). From a more theoretical perspective, Séraphin (2020) came up with a few conclusions:

- 'Planned, live events, both personal and societal in scale and meaning, will always be a prominent feature of civilization, in all societies and cultures. However, in time of pandemics or other crisis, live events will be perturbed, before a staggered returned to normal under certain conditions, and strict controls' (Séraphin, 2020: 13).

- 'As travel and tourism collapse, possibly because of the cascading effects of global warming, another energy crisis, war terrorism or global pandemics, events will also collapse temporarily, but will remain important globally because they meet fundamental human needs' (Séraphin, 2020: 14).
- 'The event professionals of the future will be advocate and competent when it comes to customer engagement during and after crisis (such as energy crisis, war terrorism or global pandemics)' (Séraphin, 2020: 26).

It is worth mentioning that the effects of the pandemics, and the way it has been dealt with, vary according to the destination.

Destination Specific

Despite the fact that travel was banned even at a national level (Jamal & Budke, 2020; Yang et al., 2020), in major cities worldwide it did not stop mobility completely. It actually triggered an exodus from major cities to countryside or coastal areas (Séraphin & Dosquet, 2020). Taking the example of France, its capital city Paris had lost 11% of its residents who had moved to their second home in the countryside/coastal area (INSEE, 2020; *Letelegramme* [Online]). In France, COVID-19 is associated with a boom in second home tourism and mountain tourism (Séraphin & Dosquet, 2020). Mountain tourism as a type of niche tourism is growing (Beedie & Hudson, 2003), but it is also a threat for the environment as most of the time it takes place in protected areas and fragile ecosystems (Geneletti & Dawa, 2009). Having said that, this type of tourism is important for the survival of rural areas, which suffer from the low density of population and lack of connection with the economic hubs (Luthe et al., 2012).

As for second home tourism, which is particularly booming in North America, Western Europe and Nordic countries (Nouza et al., 2015), it is associated with terms such as routine and novelty, inversion, back-to-nature, identity, surety, continuity, work, elitism, aspiration, time and distance, vacation home, holiday home, leisure home, etc. (Jaakson,

1986; Roca, 2013). Similar to other forms of tourism, it has benefits such as contributing to capital inflows through employment creation and income generation for small businesses (Hoogendoorn & Visser, 2015; Larsson & Muller, 2017; Visser, 2008) but also there are negative impacts such as increase in property price, crime rate, accidents, etc. (Hoogendoorn & Visser, 2015; Larsson & Muller, 2017; Visser, 2008). These negative influences are reminders of the impacts of overtourism on destinations (Séraphin et al., 2018).

Overtourism

It could be assumed that the breakout of COVID-19 would have put an end to overtourism and yet it is not the case, according to Séraphin and Dosquet (2021) who argue that with the breakout of COVID-19 and the exodus of people to countryside or coastal areas, there are some strong correlations with the characteristics of overtourism, as centred on the same four core elements, namely: tourists, locals, destinations and destination management. Equally important, Séraphin and Dosquet (2021) claim that the exodus generated by crisis is a mutation of overtourism generated by leisure purposes, marketing, low cost of tourism products and service and poor planning because (1) the motivation is survival in a pleasant and safe environment; (2) the destination is chosen according to the location of the second home; (3) tourismphobia is not towards foreigners but towards fellow citizens; (4) this new form of overtourism suggests a return to nature away from cities.

From this paragraph and preceded one, a few key points have to be reiterated, as outlined below.

Key Points from the Literature Review

This literature review highlights that for every positive impact generated by the tourism industry, there is a negative impact (Sanchez & Adams, 2008). The literature also stresses the resilience of the tourism and its ability to mutate into something else when threatened by external factors such as pandemics (in the case of this study, COVID-19). Taking the example of different businesses in the tourism sector (and related sectors), the impacts of the pandemic are not homogenous. Some destinations in France managed to make up their loss during the summer holiday break 2020, and in some cases, they exceeded their target. Having said that, all agree that summer 2021, will play a major role in the future of their business (Capital [Online]). Some organizations, such as the Sanctuary (body in charge of Lourdes Pilgrimages), adopted a radical change in their management approach as they delivered the event online via social media, which had never happened before (Lefigaro [Online]). In the following sections, this study will compare the key point of the literature review with what is happening in the luxury yachting sector. The following sections will address three research questions: (1) What are the benefits and limitations of this sector? (2) How resilient is this sector? (3) And finally, how is this sector coping in the global pandemic context?

Luxury Yachting

Luxury Products and Services

A luxury product and/or service could be defined as a product and/or service that offers customers the opportunity to experience hedonic well-being experience in very high privacy (Lloranta, 2019). From an economic point of view, the luxury segment which is worth hundreds of billion USDs worldwide (Chen & Peng, 2014) is rapidly expanding, while emphasizing the inequalities between those who have and the others (Thurlow & Jaworski, 2012). This is further supported by Correira et al. (2020) who argue that luxury has a more socially driven meaning than a self-constructed meaning. The socially driven meaning also implies for those belonging to that group to brand themselves as members of the high-end group. Indeed, luxury is also a branding strategy and positioning adopted by some destinations such as Dubai, with iconic luxury hotels such as the Burj Khalifa and Burj Arab (Heyes & Nadkarni, 2020). In the hospitality sector, hotels are classified as follows: economy, midscale, upper midscale, upscale, upper upscale and, finally, luxury (Bowie et al., 2017).

Owing to their exclusivity, luxury products and services, and particularly in the case of the hospitality sector, are often seen as not contributing to local economies, as based on the enclave model (Brenner & Aguilar, 2010). Having said that, luxury products and services in the tourism industry and cognates sectors (such as hotels), are also involved in sustainability initiatives (Sloan et al., 2013), for instance by empowering local indigenous groups so that they can maintain their legacy of local craftsmanship (Poelina & Nordensvard, 2018). This form of tourism is referred to as 'sustainable luxury tourism' (Poelina & Nordensvard, 2018). Additionally, the challenge faced by the luxury segment to remain elitist, distinctive and exclusive (Nyeck, 2004) while contributing to the local communities (Poelina & Nordensvard 2018; Sloan et al., 2013) is questioned by the existence and development of this form of tourism: 'How can we explain the resilience and growth of the luxury tourist industry at a time when climate change is acknowledged as scientific truth, and extreme weather events appear to be increasing in both frequency and devastation, while the wealth divide between the global North and the South is increasingly pronounced' (Smith, 2019: 305).

Luxury Yachting Tourism

As a form of tourism, luxury yachting tourism could be associated with nautical tourism and cruise tourism. However, if a comparison was to be made with the hospitality sector (preceding section), all other forms of nautical activities would range from economy, midscale, upper midscale, upscale and upper upscale. Luxury yachting would logically be in the upper scale category. Luxury yachting also follows the codes discussed earlier. As academic research in the field of luxury yachting tourism is very limited (Mikulic et al., 2014), this chapter uses visual research methods, employing strategies suggested by Rakic and Chambers (2012). Adopting this research method places this study in line within the trend of visual research method, as this research method is increasing across a wide range of disciplines (Rakic & Chambers, 2012). This approach is all the more relevant to this study as 'much of tourism is about image'

(Rakic & Chambers, 2012: 4). YouTube will be the primary source of information as films (videos) can contribute to new understandings of tourism (Rakic & Chambers, 2012: 4). This approach is appropriate as Jamal and Budke (2020) explain that when a topic (such as COVID-19) is unfolded, it is allowed to use media sources. The topic of this paper falls in the field of unfolding topics (Mikulic et al., 2014). As a matter of fact, to discuss the impacts of COVID-19 on the event and tourism industry, Séraphin (2020) uses YouTube as part of a research approach based on bricolage. This study applies the same framework as used by Séraphin (2020) to collect data on how practitioners foresee the future of the tourism industry in a COVID-19 context. Specifically, two YouTube videos have been used. The first one is on luxury products and services, and all the activities and businesses which are gravitating around the luxury sector. The second one is on mega-yacht. Detailed information about the framework and information collected can be found in Appendix 1.

Based on Appendix 1, luxury yachting is seen as more than a form of tourism, but a driver for tourism development. Indeed, taking the example of Montenegro, yachting has led to the development of a marina, which has in turn led to the development and revival of local villages. The case of Montenegro shows that there is a whole economic ecosystem around yachting, as summarized in Figure 2.1.

It is also worth highlighting that yachting is also a lifestyle for some, as it is a way for the 1% of the population who can afford a luxury yacht, to show off their success (professionally), and subsequently, their wealth. For some, it is just a way to be happy and make their friends and

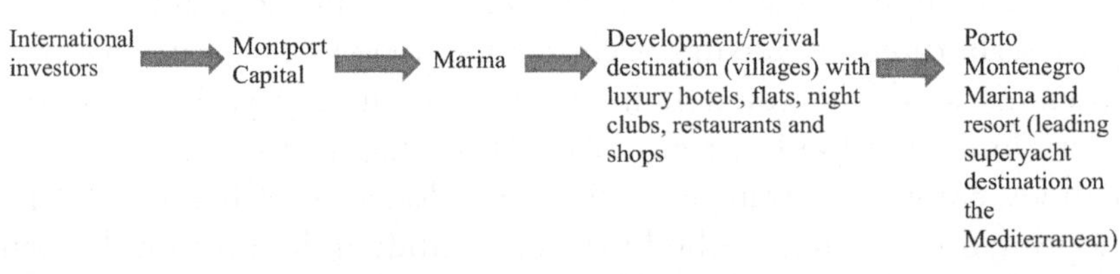

Fig. 2.1 Montenegro development model around luxury yachting tourism (*Source* The author)

relatives happy. There is a large number of people who also benefit from this sector. Among these are crew, designers, architects, etc.

An economic and non-economic dimension can be identified in yachting. This industry could therefore be said to be ambidextrous (Figure 2.2).

Despite the fact that luxury products and services cultivate their differences compared with more affordable ones, it appears that they also follow some trends, such as the protection of the environment. Indeed, superyachts follow the trend towards sustainability in place in all economic sectors (Visser, 2015), but also in the tourism industry (Sloan et al., 2013). To do so, they use, for instance, carbon neutral fuels such as hydrogen and ammonia (Eastlack et al., 2019).

With respect to research question 1, the discussion above demonstrates that there are benefits and limitations of the luxury yachting sector. Firstly, at an individual/micro level, luxury yachting contributes to the happiness and quality of life of those who can afford this type of product and service. Secondly, at a wider community/macro level, this sector creates an ecosystem of employment and development for some local

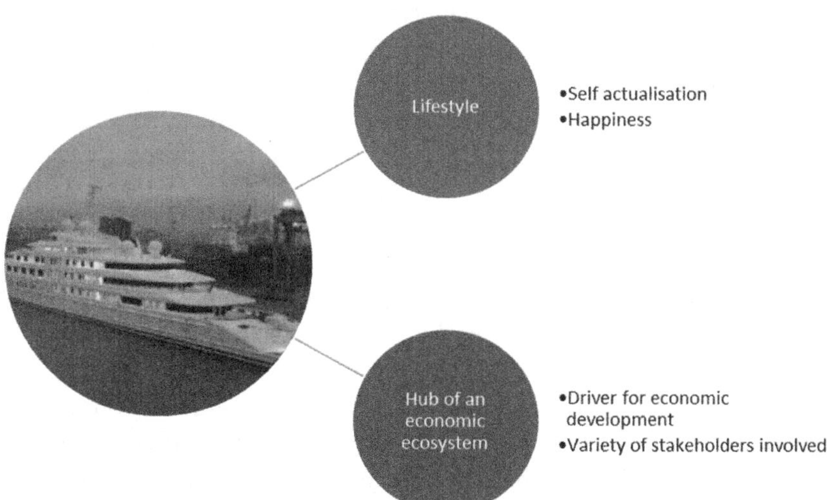

Fig. 2.2 Ambidextrous nature of yachting (*Source* The author)

communities. The question we raise is how has the reality of the current global pandemic affected the industry?

Luxury Yachting Tourism in a COVID-19 Context

As any other sector, yachting follows hygiene protocol (boatinternational.com [Online]). The breakout of the virus has not impacted (directly) the sector much due to the fact that:

> *A yacht offers an easily controlled environment in terms of the comings and goings of guests and other personnel. The passerelle is effectively a drawbridge and nobody need step on board the yacht as deliveries are left on the quay and loaded by the permanent crew'*. (boatinternational.com [Online])

> *Private charter ensures guests can stay safeguarded in smaller, immaculately-clean environments, while minimising the need to frequent public areas where risks naturally increase.* (boatinternational.com [Online])

The above quotes present yachts/yachting as a self-contained environment. Luxury yachts could even be compared to moving islands. For Korstanje (2018), islands are compounds and utopian paradise. Korstanje (2018: 79) further adds that they are 'home as a secure place'. On that basis, yachting could be said to be able to impact positively the ecosystem which it is part of. This allows us to address research questions 2 and 3, respectively: How resilient is this sector? How is this sector coping in this global pandemic context? Destinations do not benefit as much from this sector, not because yachts are docked but due to the fact that restrictions have impacted the opening of natural and built attractions, meaning that yacht owners (and crew) cannot visit, and/or buy groceries (and other supplies) as usual (Superyachttimes.com [Online]). However, what might decrease rather significantly in the future is the number of superyachts to be bought. Indeed, in times of crises and/or austerity consumers rethink their spending priorities (Savelli, 2011). The negative impacts of the pandemic on luxury yachting have mainly been indirect as many superyacht events have been cancelled such as the *Cannes Yachting Festival*, the *Southampton Boat Show*, *Rolex Swan Cup*, *Monaco Yacht Show*, etc. Others have been postponed such as the *Australian Superyacht Rendezvous* and *Perini Navi Cup* (boatinternational.com [Online]).

Conclusion

As a component of the tourism industry, luxury yachting has been affected by the breakout of COVID-19. However, as opposed to other sectors of the industry, the negative impacts on yachting itself has not been major. The major negative impacts have been for the related sectors and services. A yacht could therefore be considered as a safe 'bubble' within an uncertain world. The key features of luxury, and therefore yachting, namely privacy, exclusion, exclusivity, etc., quite often pointed out as negative, nevertheless in times of crises, such as pandemics, they appear to be assets. When designing tourism products and services, it is therefore important to bear in mind a Janusian thinking approach, which enables the design of products and services that can be used regardless of the context, as proven by the luxury yachting sector.

Appendix 1: Luxury Yachting (YouTube Analysis)

Item No	Authorship	Date released/ date published	Source	Title	Audience	Production[a]	Context/ frame[b]	Summary/key points
1	Capital M6	04.6.17	YouTube https://https://www.youtube.com/watch?v=UO9y6DCwFo4	Villas de luxe, destinations de rêve. Le business des vacances sans limite	General public	Pre-COVID	Programme on luxury yacht owners	Since 2007, Montenegro new destination for rich people, as other popular spots (St Tropez, etc.) are too expensive, noisy, etc Yacht (15 million euros/42 metres long/ 5 bedrooms/marble,etc.) 100.000 euros per year to moor yacht in marina (south of France). In Montenegro, less than ½ price. Per year, it costs 67.000 euro for a 50-metre-long yacht; 230.000 euro for 100-metre-long yacht. VAT is only 7% (3 times less than in France). The marina is also very safe (368 CCTV). Fuel is free of tax (45% cheaper than other destinations part of the E.U). As a result, marina is fully booked all year-round. Booking needs to be done 3 months before at least

(continued)

(continued)

Item No	Authorship	Date released/ date published	Source	Title	Audience	Production[a]	Context/ frame[b]	Summary/key points
								Montenegro is developing luxury tourism Peter Munk, Bernard Arnault, and Jacob Rothschild, set up a company called 'Montport capital' and invested 402 million euros is the development of a marina for yacht (50–250-metre-long), with 450 berths available (Monaco's marina is twice bigger) Prince of Saudi for instance own yacht A village has been built around the marina to encourage yacht owner to stay longer. Within the marina can be found: 228 luxury flats/a 5* hotel/1 nightclub/55 restaurants and luxury shops (managed by Montport Capital)

(continued)

(continued)

Item No	Authorship	Date released/ date published	Source	Title	Audience	Production[a]	Context/ frame[b]	Summary/key points
2	Imineo	21.11.19	https://https:// www.you tube.com/ watch?v=lvl ozrSklOc	Méga-yachts: Enquête sur la nouvelle passion des milliardaires	General public	Pre-COVID	Programme on luxury yacht owners	Yachts are getting bigger and bigger The industry is involving Key players are: Arabic Sheiks, Russians and American businessmen. There is a competition between them (who will get the bigger and more expensive yacht) The most expensive yacht is worth 550 million euros (belongs to Abu Dhabi Sheik) The owners are billionaires. It is a small and close group (1% population), with their rituals and secrets Designers like Phillip Stark have made a name for themselves. There is an entire economy around yachting (e.g. Yearly paid staff, etc.) Luxury yachting is a lifestyle, pleasure, a passion, 2nd home, a way to show off wealth, etc

(continued)

(continued)

Item No	Authorship	Date released/ date published	Source	Title	Audience	Production[a]	Context/ frame[b]	Summary/key points
								Monaco is the place with the most billionaire per km2, and a hot spot for superyacht 360 companies in Monaco have a connection directly or indirectly with yachting (design companies, etc.) 2003: 1st mega-yacht in the world (owned by Paul Allen) Yachting salon of Monaco/Boat show Dubai

Source The author (Adapted from Hammond & Wellington, 2013; Séraphin, 2020)

[a]What were the social, political and cultural conditions in which it was produced?

[b]Whose views of events have been recorded in documentary format?

References

Beedie, P., & Hudson, S. (2003). Emergence of mountain-based adventure tourism. *Annals of Tourism Research, 30*(3), 625–643.

Boat International. (2020). Retrieved September 24, 2020, from https://www.boatinternational.com/yachts/news/coronavirus-superyacht-events-cancelled--42933

Bowie, D., Buttle, F., Brookes, M., & Mariussen, A. (2017). *Hospitality marketing* (3rd ed.). Routledge.

Brenner, L., & Aguilar, G. (2010). Luxury tourism and regional economic development in Mexico. *The Professional Geographer, 54*(4), 500–520.

Capital M6. (2017, June 4). *Villas de luxe, destinations de rêve. Le business des vacances sans limite*. Retrieved September 24, 2020, from https://www.youtube.com/watch?v=UO9y6DCwFo4

Chen, A., & Peng, N. (2014). Examining Chinese consumers' luxury hotel staying behaviour. *International Journal of Hospitality Management, 39*, 53–56.

Correira, A., Kozak, M., & Del Chiappa, G. (2020). Examining the meaning of luxury in tourism: A mixed-method approach. *Current Issues in Tourism, 23*(8), 952–970.

Eastlack, E. et al. (2019). Zero emission super-yacht. In *2019 Fourteenth International Conference on Ecological Vehicles and Renewable Energies (EVER)* (pp. 1–8), Monte-Carlo, Monaco.

Geneletti, D., & Dawa, D. (2009). Environmental impact assessment of mountain tourism in developing regions: A study in Ladakh, Indian Himalaya. *Environmental Impact Assessment Review*, doi: https://doi.org/10.1016/J.EIAR.2009.01.003

Getz, D., & Page, S. J. (2020). *Events studies: Theory, research and policy for planned events*. Routledge

Hammond, M., & Wellington, J. (2013). *Research methods. The key concepts*. Routledge.

Heyes, A., & Nadkarni, S. (2020). Luxury consumption in tourism: The case of Dubai-Part 2. *Research in Hospitality Management, 10*(1), 51–53.

Hoogendoorn, G., & Visser, G. (2015). The economic impact of second home development in small-town south Africa. *Tourism Recreation Research, 35*(1), 55–66.

Imeo. (2019, November 21). *Méga-yachts: Enquête sur la nouvelle passion des milliardaires.* Retrieved September 24, 2020, from https://www.youtube.com/watch?v=IvlozrSkIOc

Insee. (2020). *Population présente sur le territoire avant et après le début du confinement–Premiers résultats.* Retrieved September 24, 2020, from www.insee.fr/fr/information/4477356/

Jaakson, R. (1986). Second-home domestic tourism. *Annals of Tourism Research, 13*(3), 367–391.

Jamal, T., & Budke, C. (2020). Tourism in a world with pandemics: Local-global responsibility and action. *Journal of Tourism Futures.* https://doi.org/10.1108/JTF-02-2020-0014

Korstanje, M. (2018). *The mobilities paradox: A critical analysis.* Edward Elgar

Larsson, L., & Muller, D. K. (2017). Coping with second home tourism: Responses and strategies of private and public service providers in western Sweden. *Current Issues in Tourism, 22*(16), 1958–1974.

Le Télégramme. (2020, April 28). *Déconfinement, ce qu'il faut retenir du plan d'E. Philippe.* Retrieved September 24, 2020, from https://www.letelegramme.fr/france/deconfinement-par-departement-ecoles-commerces-ce-qu-il-faut-retenir-du-plan-d-edouard-philippe-28-04-2020-12545141.php

Lloranta, R. (2019). Luxury tourism service provision: Lessons from the industry. *Tourism Management Perspectives, 32,* 100568. https://doi.org/10.1016/j.tmp.2019.100568

Luthe, T., Wyss, R., & Schuckert, M. (2012). Network governance and regional resilience to climate change: Empirical evidence from mountain tourism communities in the Swiss Gotthard region. *Regional Environment Change, 12,* 839–854.

Mikulic, J., Kresic, D., & Kozic, I. (2014). Critical factors of the maritime yachting tourism experience: An impact-asymmetry analysis of principal components. *Journal of Travel & Tourism Marketing, 32*(1), 30–41.

Mihalache, M., & Mihalache, O. R. (2016). Organizational ambidexterity and sustained performance in the tourism industry. *Annals of Tourism Research, 56,* 142–144.

Nouza, M., Olafsdottir, R., & Saelporsdottir, A. D. (2015). Motives and behaviour of second home owners in Iceland reflected by place attachment. *Current Issues in Tourism, 21*(2), 225–242.

Nyeck, S. (2004). Luxury brands online and offline: The case of French brands. *European Retail Digest, 41,* 1–5.

Poelina, A., & Nordensvard, J. (2018). Sustainable luxury tourism, indigenous communities and governance. In M. Gardetti & S. Muthu (Eds.), *Sustainable luxury, entrepreneurship, and innovation: Environmental footprints and eco-design of products and processes* (pp. 147–166). Springer

Rakic, T., & Chambers, D. (2012). *An introduction to visual research methods in tourism*. Routledge.

Roca, Z. (2013). *Second home tourism in Europe: Lifestyle issues and policy responses*. Farnham.

Sanchez, P. M., & Adams, K. M. (2008). The Janus-faced character of tourism in Cuba. *Annals of Tourism Research, 35*(1), 27–46.

Savelli, E. (2011). Role of brand management of the luxury fashion brand in the global economic crisis: A case study of Aeffe Group. *Journal of Global Fashion Marketing, 2*(3), 170–179.

Séraphin, H. (2020). COVID-19: An opportunity to review existing grounded theories in event studies. *Journal of Convention and Event Tourism*. https://doi.org/10.1080/15470148.2020.1776657

Séraphin, H. (2021). COVID-19 and the acknowledgement of children as stakeholders of the tourism industry. *Anatolia*. https://doi.org/10.1080/13032917.2020.1856690

Séraphin, H., & Butcher, J. (2018). Tourism management in the Caribbean: The case of Haiti. *Caribbean Quarterly, 64*(2), 254–283.

Séraphin, H., & Dosquet, F. (2020). Mountain tourism and second home tourism as post COVID-19 lockdown placebo? *Worldwide Hospitality and Tourism Themes*. https://doi.org/10.1108/WHATT-05-2020-0027

Séraphin, H., & Dosquet, F. (2021). COVID-19 and the mutation of over-tourism. The case of France. In H. Séraphin, & A. Yallop (Eds), *Overtourism and tourism education: A strategy for sustainable tourism futures*. Routledge

Séraphin, H., Sheeran, P., & Pilato, M. (2018). Over-tourism and the fall of Venice as a destination. *Journal of Destination Marketing & Management, 9*, 374–376.

Séraphin, H., & Yallop, A. (2019). Proposed framework for the management of resorts mini clubs: An ambidextrous approach. *Leisure Studies, 38*(4), 535–547.

Sharma, A., & Nicolau, J. L. (2020). An open market valuation of the effects of COVID-19 on the travel and tourism industry. *Annals of Tourism Research*. https://doi.org/10.1016/j.annals.2020.102990

Sigala, M. (2020). Tourism and COVID-19: Impacts and implications for advancing and resetting industry and research. *Journal of Business Research, 17*, 312–321.

Sloan, P., Legrand, W., & Chen, J. S. (2013). *Sustainability in the hospitality industry: Principles of sustainable operations* (2nd ed.). Routledge.

Smith, O. (2019) Luxury, tourism and harm: A deviant leisure perspective. In T. Raymen, & O. Smith (Eds.), *Deviant leisure: Palgrave studies in crime, media and culture* (pp. 17–44). Palgrave Macmillan.

Superyachttimes. (2020). Retrieved September 24, 2020, from https://www.superyachttimes.com/yacht-news/coronavirus-superyacht-industry-updates

Thurlow, C., & Jaworski, A. (2012). Elite mobilities: The semiotic landscapes of luxury privilege. *Social Semiotics, 22*(4), 487–516.

Tiwari, P., Séraphin, H., & Chowdhary, N. (2020). Impacts of covid-19 on tourism education: Analysis and perspectives. *Journal of Teaching Travel and Tourism.* https://doi.org/10.1080/15313220.2020.1850392

Visser, G. (2008). South Africa has second homes too! An exploration of the unexplored. *Current Issues in Tourism, 3*(4–5), 351–383.

Visser, W. (2015). *Sustainable Frontiers: Unlocking change through business, leadership and innovation.* Sheffield: Greenleaf Publishing.

Vo-Thanh, T., Séraphin, H., Okumus, F., & Koseoglu, M. A. (2020). Organizational ambidexterity in tourism research: A systematic review. *Tourism Analysis, 25*(1), 137–152.

Yang, Y., Zhang, H., & Chen, X. (2020). Coronavirus pandemic and tourism: Dynamic stochastic general equilibrium modelling of infectious disease outbreak. *Annals of Tourism Research, 83.* 102913, https://doi.org/10.1016/j.annals.2020.102913

3

Understanding the Luxury Charter Yacht Value Proposition: Co-creating Customer Experiences

Lyn Batchelor and Russell Williams

Introduction

The luxury yacht charter market exists within the wider luxury and luxury tourism markets. In 2017 the luxury yacht market size was valued at \$5703.4 million and is expected to reach \$10,205.7 million by 2025, a Compound Annual Growth Rate (CAGR) of 7.8% from 2018 to 2025 (Allied Market Research, 2019). The size and growth of luxury yachting make it an attractive proposition for firms and provides a strong rationale for understanding both its characteristics and the motivations of consumers towards it. Indeed, understanding luxury enables firms,

L. Batchelor · R. Williams (✉)
The University of Aberdeen, Qatar Campus, Doha, Qatar
e-mail: Russell.williams@abdn.ac.uk

L. Batchelor
e-mail: Lyn.Batchelor@abdn.ac.uk

T. Gladkikh et al. (eds.), *Luxury Yachting*,
https://doi.org/10.1007/978-3-030-86406-4_3

including those in the luxury yacht charter space, to design, deliver and communicate more competitive value propositions.

To understand the luxury yacht charter market this chapter provides some context in terms of data for the size and growth of luxury, luxury tourism and the luxury yacht market. Following this overview, academic interest and insight into how luxury and luxury tourism is conceptualized are detailed. This academic overview provides a useful framework for approaching and understanding what it is that luxury yacht charter consumers buy. Importantly, this chapter also details an emerging trend in luxury consumption: a shift in focus away from the 'material' to the 'experiential' (Yang & Mattila, 2017), or what has been labelled the 'experience economy' (Pine & Gilmore, 1998). Using the 4Es framework of Pine and Gilmore (1998, 1999) and reflecting on a luxury yacht charter experience, this chapter provides an illustrative inventory of each of these 4Es. This chapter concludes with a reflection on how luxury yacht charter firms can deliver a more competitive offer through an understanding of the importance of experience in value creation.

The Attractiveness of the Luxury (Tourism) Market

Luxury consumption has increased significantly in recent times making it an important and potentially attractive part of the market for firms to target. Bain and Company (in D'Arpizio et al., 2017) note the luxury market tripled in size between 1994 and 2015, with the 2017 total spend estimated at 1160 Bn Euros. Within this estimated spend, hotel and exclusive vacations (of which the luxury yacht charter market is a part) make up 193 Bn Euros. Importantly, while recent growth in luxury consumption has been considerable, it is set to grow still further. This future growth is in part due to an increase in the number of consumers having access to luxury through increasing wealth and a reduction of personal budget constraints (a so-called 'democratisation' of luxury meaning luxury is no longer the domain of an exclusive elite) (Yeoman, 2011, 2012; Yeoman & McMahon-Beattie, 2011).

Illustrating the growth in luxury, a Boston Consulting Group (BCG) forecast for the luxury segment reports that the total number of luxury consumers is estimated to increase from 400 million in 2015 to 480 million by 2022 and that this is from all types of consumers: 'new aspirational consumers' trading up to luxury goods at 365–435 million; 'top aspirational consumers' at 20–26 million; and 'top luxury consumers at 16–20 million consumers' (cited from Pianon et al., 2017). Other data confirm the growth in those likely to purchase luxury. For example, Credit Suisse (2018) identifies 42.2 million $ millionaires worldwide, an increase of 2.3 million over the previous 12 months. In terms of geography, the United States added 878,000 new millionaires (~40% of the global increase) with France, Germany, the United Kingdom and Italy adding around 200,000 each. In China, the number of millionaires rose by a modest 186,000.

Two other factors potentially make the luxury tourism segment an attractive one for firms to target and serve. First, a relatively small number of individuals deliver disproportionate spending. Xenia in Trihas et al. (2020) for example reports that luxury tourists account for only 3% of total tourist arrivals while contributing 25% of tourist revenue. Second, luxury and luxury tourism appear largely unaffected by fluctuations in the global economy (Trihas et al., 2020). Moreover, industry reports have indicated a significant upturn in interest in yacht chartering because of the COVID-19 pandemic as they provide an opportunity for consumers to holiday in a less risky manner through controlled, more private and less crowded environments (Froehlich, 2020).

One final note on the luxury market is that its location is shifting. Whilst the luxury segment (the consumers) has, historically, been the domain of consumers from more mature and developed economies (i.e. consumers from North America, Japan and European countries), there is now increasing focus elsewhere on the globe (Kapferer & Valette-Florence, 2016). For the moment though the pandemic has appeared to press pause on the 2020 superyacht order book as shipyard labour was not available and travel restricted (Boat International, 2021). Arguably, the effect of China's recent anti-corruption stance over the last few years and Brexit has also tempered the demand for conspicuous (yacht) consumption (ibid.).

The Luxury Yacht Charter Market

The word yacht derives from the Dutch word 'jacht', meaning a vessel sailed for pleasure or to carry important people. Spence (2016: 318) defines the luxury domain: '*Superyachts are luxury privately owned, and professionally crewed vessels over 30 metres in length. An average motor yacht measures around 47 metres in length and costs approximately €30 million to buy new from a reputable European shipyard*'. In addition to build and purchase costs, an '*owner should expect to spend about 10% of the yacht's initial cost on annual upkeep*' (Adamczyk, 2015: para. 5) with crew wages lifting this to millions per annum. *Therefore, 'superyachting is an exclusive leisure practice reserved for the superrich*' (Spence, 2016: 318).

Beyond this overview (average) perspective, the industry can be seen to segment the market through yacht size, type and application, as well as region (Jadhav & Singh, 2020). For size, this means large (over 50 m), medium (30 m–50 m) and small (up to 30 m). For type, this means sailing yachts, motor yachts, and others (inflatable boat, Catamaran, gullet). In the yacht market application means vacation/leisure, sailing, and 'other' categories. Finally, for regions, the market is North Americas, Europe, Asia–Pacific, and Latin America, Middle East, and Africa.

Moreover, luxury yachting is growing. Indeed, Deloitte Italy (2019) reported 160 superyacht units delivered in 2018 (up 19, 2.9% in 2017). Delays have been reported in delivering on the 2021 global superyacht order book of 821 superyacht projects (Boat International, 2021) although project delays are not unusual as COVID has severely interrupted delivery schedules.

At the firm level, some of the key players in the charter market are Sailogy S.A., Yachtico Inc., Zizooboats GmbH, Fraser Yacht, Boatsetter, Northrop and Johnson, Nautal, Martello Yachting and Company, Princess Yacht limited and Charter Yachts Australia.

As a point of clarification, what differentiates luxury yachting from luxury cruising is that: '*...the mobility of the yacht is largely determined by the whimsical demands of its super-rich passengers (hereafter guests). In addition, the ratio of crew members to guests onboard a yacht is close to 1:1* *(depending on the length and size of the vessel) which enables a more personal and attentive service to guests and which thus creates different social-spatial interactions compared to large passenger vessels such as cruise ships*' (Spence, 2014: 402).

Understanding Luxury

Mirroring the growing presence of luxury in the economy, due attention has been given to it by scholars from a variety of different discipline areas (Kapferer & Valette-Florence, 2016). Indeed, in a recent 2000–2015 systematic review on the ISI Web of Knowledge database Gurzki and Woisetschläger (2017) retrieved 1315 records from 2425 authors published in 533 journals across 89 different research fields. The most active of these fields were: 'Business and Economics' (849 publications); 'Business' (457); 'Economics' (310); 'Social Sciences – other topics' (202); 'Psychology' (171); 'Management' (160); 'Hospitality' (112); 'History' (95); 'Sociology', (72); and 'Social Sciences' (65). This range of interest helps explain why luxury is notoriously difficult to define (Tynan et al., 2010; Vigneron & Johnson, 1999). However, a logical starting point for the design and development of a competitive luxury yacht value proposition is an interpretation of what luxury is. So, what is luxury? What characteristics does luxury possess? What motivations do individuals have towards it? Moreover, what is the major trend in luxury that firms need to be cognisant of as they look to serve this market?

A starting point for advancing an understanding is the word luxury itself. Luxury derives from the Latin term *luxux* and means sensuality, splendour and pomp. Dictionary.com (2021) defines luxury as 'a material object, service, etc., conducive to sumptuous living, usually a delicacy, elegance, or refinement of living rather than a necessity'. But luxury appears to be something more than this, as considerable academic literature reveals.

From among many academic articles, the following contributions illustrate some of the suggested perceived *characteristics* associated with luxury: beauty, uniqueness, creativity, sensuality and exclusiveness (Kapferer, 1998); exclusivity and high quality (Phau & Prendergast,

2000); and unique design, superb performance or durability and extraordinary superiority to comparable substitutes which justify its high price (Mandhachitara & Lockshin, 2004). These three works alone illustrate some of the variations in defining the characteristics or perceptions of luxury.

Going beyond individual works, additional insight into luxury can be obtained from other, more recent, systematic reviews. For example, Ko et al. (2019) report five key dimensions associated with consumers' perception of a *luxury brand;* be of high quality; offer authentic value via functional and emotional benefits; a prestigious image built on the qualities of artisanship, craftsmanship or service quality; be worthy of obtaining a premium price, and be capable of inspiring deep connection, or resonance with the consumer. The inference here is that luxury branding is associated with the *characteristics* of the goods (and/or services). Importantly, some of these characteristics can be understood as comprising tangible attributes (for example, price, quality, beauty, craftsmanship, exclusiveness) and, additionally, intangible and symbolic attributes (for example, perceived authenticity and uniqueness) (Turunen & Laaksonen, 2011).

Employing the search terms 'luxury goods' and 'consumer behaviour' or 'Factors' or 'Attributes' or 'consumption' or 'purchase' for ranked journals, the systematic review of Dhaliwal et al. (2020) concluded that the variables influencing consumer behaviour of luxury buying fall into four categories: (1) personal; (2) psychological; (3) cultural and social; and (4) factors related to luxury goods. These categories indicate aspects of luxury go beyond a characteristics approach (i.e. something inherent in the item). Importantly, they provide insight that luxury has individual (subjective) as well as interpersonal dimensions. That is, luxury does not exist as something separate from an individual or from the individual as a part of a social group. The consequence of this is that luxury for one person may not be luxury for another. This is perhaps most easily illustrated when considering (a high) price, a factor often associated with luxury. If luxury is associated with high price by its potential role in determining or signalling exclusivity, its impact is only really apparent when also considered alongside an individual's income, budget or wealth (Yeoman & McMahon-Beattie, 2006).

The individual and interpersonal dimensions of luxury are also highlighted from a behaviourist perspective built on two types of motivation: intrinsic and extrinsic (Skinner, 1953). Intrinsic motivations are those motivations that come from within and are about the individual self. In terms of luxury, they may be related to the value an individual derives from self-fulfilment and pleasure. Conversely, extrinsic motivations are external motivations, so, for example, value may be derived from that obtained from signalling something to others. This might be status from being seen alongside a luxury item. Luxury is derived from 'seeing and being seen' (Mortelmans, 2005), where seeing is the search for signals derived from luxury that can subsequently be used to 'be seen'. Being 'seen' highlights the signalling role or motivation for luxury, recognized in the seminal works of Thorstein Veblen and his reference to conspicuous consumption, snob, and bandwagon effects. For Veblen (1899) conspicuous consumption refers to purchase motivations based on the desire to display wealth, status and power and, importantly, to achieve a sense of belonging to a reference group. Signalling luxury purchases provides a sense of belonging to an elite group and makes its rationale more public (social) than private (internal). The snob effect describes how demand for a good/service increases as its quantity is decreased. If a high price of an item creates exclusivity it will create a preference for the item. Conversely, the bandwagon effect works in the opposite direction. Demand increases when others are seen to consume it. Both effects are prevalent in the luxury market (Vigneron & Johnson, 1999).

The public or social motivation aspect for conspicuous luxury consumption has largely followed an economics tradition where it has been associated with the desire to signal or demonstrate status or wealth to reference groups. However, publicly visible conspicuous consumption is also associated with signalling *who* you are, for example, your taste or your self-image to reference groups (Chen et al., 2008). This is about signalling personality and identity construction and demonstration (Bronner & de Hoog, 2019). Of course, publicly visible consumption is not to be read as being derived only from tangible items. Signalling can come from non-tangibles, including vacations.

A Key Trend for Luxury

For Boven and Gilovich (2003) consumption purchases exist on a continuum with material purchases at one end and experiential purchases at the other. Applying a slightly more nuanced model for luxury branding, Berthon et al. (2009) present a three-dimensional value model with functional, symbolic and experiential value dimensions: Functional value relates to the physical attributes of the product and what the brand does (e.g. material quality, craftsmanship, durability, etc.); Symbolic value is related to what the item and its brand signals to others and the value of that signalling to the signaller (e.g. the duck egg blue of Tiffany box); Experiential value relates to what the product (and brand) means to the individual (e.g. the sensations, emotions, thoughts, etc.). Importantly, the 'centre' or 'locus' of value is changing. Whereas the locus was once focused on the functional dimension and then the symbolic, there is now a shift towards what, following Pine and Gilmore (1998, 1999), has widely come to be known as the 'experience economy' (ibid.). It is a shift away from the material (Yang & Mattila, 2017), one already beginning to be noted by the luxury industry (CBI, Ministry of Foreign Affairs, 2018).

To understand the 'experience' that makes up the experience economy and what it means for luxury charter yacht tourism, a useful starting point is a definition. The Cambridge Dictionary (2021) defines experience as: 'something that happens to you that affects how you feel'. In contrast to goods and services which are external to the customer, experiences are internal, personal and exist only in the mind of the consumer who has been engaged on some level, be it physical (appealing to our senses), or intellectual, emotional and spiritual (Hollebeek et al., 2014; Pine & Gilmore, 1998; Tsaur et al., 2007). Importantly, for marketers, memorable experiences, or indeed the very anticipation of them, create beliefs and attitudes, and influence behaviours that consumers are prepared to pay for (Oh et al., 2007), and pay a premium for in the case of luxury experiences. Creating and managing customers' (unforgettable) consumption experience is, therefore, an important part of a luxury value proposition and is the basis of competitive advantage for firms.

In understanding experiences, a subtle, but important, point of difference can be drawn from services marketing. Services marketing is concerned with supply, i.e. the process or delivery-focused service economy. The focus is on a set of (service) *activities* carried out on the consumers' behalf. Services undoubtedly play an important part in luxury tourism, but they contrast to experiences in that experiences relate to a series of memorable events designed to engage the consumer in a personal way. Experience marketing is thus about taking the essence of a product/service and augmenting it as a set of experiences.

The experience model of Pine and Gilmore (1998, 1999) has been widely used across leisure and tourism contexts, e.g. heritage trails, Hayes and MacLeod (2007); bed-and-breakfast accommodation, Oh et al. (2007); cruising, Hosany and Witham (2010); festivals, Mehmetoglu and Engen (2011); retailing, Manthiou et al. (2014); casino hotels, Wang et al. (2013); temple stays, Song et al. (2015); and wine-tourism experience, Vo-Thanh and Kirova (2018). As the suggested 4Es framework stands out among applications of the experiential view of consumer behaviour (Quadri-Felitti & Fiore, 2012), it is a logical starting point for understanding the luxury yacht chart experience.

For Pine and Gilmore (1998, 1999) organizations can meet the needs of consumers by staging desirable experiences. These experiences are likely to include some combination of four categories of experience: education, entertainment, escapism and esthetic (the 4Es framework). These experiences are about learning (education), feeling (entertainment), esthetic (presence) and doing (escapism) (Hosany & Witham, 2010; Mehmetoglu & Engen, 2011; Vo-Thanh & Kirova, 2018).

Each of the experiences is classified in two dimensions: the first of these dimensions describe the level of consumer involvement or participation in the experiences, which may be *passive participation* at one end of the spectrum or *active* at the other (this dimension forms the horizontal axis of the model). With passive participation, the consumer does not influence the staged experience. Consumers watch a play or a snooker tournament from a distance. Passive participation is akin to 'gaze' in tourism. Active participation entails co-creating the experience or joining in to create an experience. Wine tourism and heritage ecotourism are well-cited examples. The second dimension of the consumer experience is classified by the connection of the consumer with events, acts or performances. Absorption exists at one end of the spectrum and immersion at the other end (this dimension forming the y axis). Absorptive experiences involve the mind, for example travelling to Provence to learn the French language. Immersive experiences are those mobilizing the whole self. Examples of immersive experiences would be feeding the animals at a zoo or participating in cultural dancing. Combined, the domains and the axis create four dimensions: education (active and absorption), entertainment (passive and absorption), aesthetics (passive and immersion) and escapism (active and immersion) (Figure 3.1). Importantly, the experiential zones are not intended to be mutually exclusive rather they are permeable and interlinked. For Pine and Gilmore (1998) the richness of the consumer's experience is likely to be influenced by how all four zones are incorporated in the experience.

Authenticity was later identified as an important element in consumer experiences (Gilmore & Pine, 2007). Authenticity is about being authentic, real, genuine, etc. A sense of authenticity includes 'genuineness', the 'real thing', 'legitimacy' and 'believability' (Di Domenico & Miller, 2012). Following Wang (1999), authenticity has three formats: object, constructed and existential. In overview, object authenticity refers to an object. Constructive authenticity refers to that projected onto

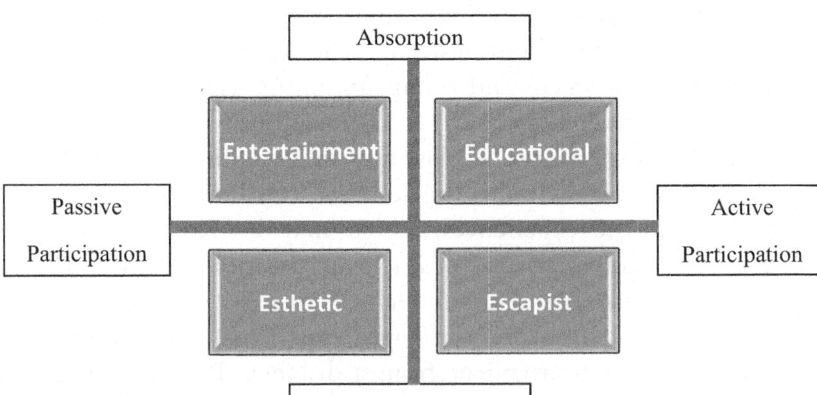

Fig. 3.1 The 4E model of customer experience

objects. Both objective and constructive authenticity are hence object-related. In the case of tourism and constructive authenticity, there may be several versions of authenticity that come from either tourists or tourist providers. Existential authenticity refers to the existential states of being that can be activated by activities. For tourism, existential authenticity may have nothing to do with the authenticity of toured objects. Importantly, authenticity has been shown to increase tourists' perceived value and satisfaction (Chen & Chen, 2010), as well as satisfaction and behavioural intentions and loyalty for cultural heritage tourism (Park et al., 2019).

Luxury Charter Yacht Tourism through the Lens of the Experience Economy

The three-dimensional value model (Berthon et al., 2009) provides a framework for understanding the luxury yacht charter market. In terms of the functional (material) dimension, aesthetics, quality of materials and quality of finish (craftsmanship) of both the yacht and the artefacts on it will all play an important part in the luxury value proposition. Following the idea noted above that material elements are subjectively interpreted (e.g. they may be defined individually by income, wealth and budget constraint, culture, etc.), it appears logical to assume that these material elements must meet minimum threshold specifications to be deemed luxury.

With the symbolic dimension, value derives from signalling to self as well as to others. Luxury value comes from either or both of 'seeing' and 'being seen'. For the individual, it is important to see themselves alongside something (or indeed somewhere—think luxury marina or remote location), which for them has luxury connotations. The 'something' or 'somewhere' provides self-actualization, the ability to create a more idealized image of the self (Danziger, 2006). In terms of being seen, the value lies in the symbolism of association that demonstrates value to others, whether these others are on the yacht with the individual, or at destinations, or even those back at home (Correia et al., 2016). Symbolic value is about associations reflecting both an individuals' own preferences and attitudes and those linked to others and how they view these associations. Together, both seeing and being seen, provide important value, supporting and defining an individual's life narrative and sensemaking (Gretzel et al., 2006).

For Berthon et al. (2009), experiential value relates to what a product, or in this case the yacht charter, means to the individual in terms of the sensations, emotions, thoughts, etc. Considering experiential value through the lens of the 4Es framework set out by Pine and Gilmore (1998, 1999), the next section draws on one of the author's of this chapter personal reflection of a luxury yacht experience to provide an 'inventory of experiences' and discussion for each of the four categories of experience.

Methodologically, the account is recounted as a partial ethnography (Grbich, 2007). In this case, the author experiences the luxury yacht charter while not being part of the tribe of the 'rich and famous', but can experience, listen and observe while immersed in the experience. The next section offers a first-person account as a lens to view the luxury charter experience.

Luxury Charter Yacht Tourism through the Lens of Personal Experience

The one-week luxury yacht experience was centred on the Whitsunday area of the Great Barrier Reef, Australia. The 2300 km long reef is globally renowned for its corals, fish, turtles and white silica sands. Typically, visitors experience the reef as a day trip (just over an hour via fast catamaran from Airlie Beach, or two hours by sailboat) engaging in sightseeing, snorkeling or diving. These commercial experiences are 'limited to' 300 people a day. Or as day trips from resort islands.

Our experience began many miles from the luxury yacht itself and involved a 1000 km business class flight. As we circled over the marina before landing, we were rewarded by the sight of the yacht. The yacht was too big to be moored inside the marina and was instead outside of the last pontoon. The luggage of our fellow flight travellers was added to a baggage train and they were taken by minibus to other resorts or marinas. We were instead met by the ship's captain in the full white dress uniform who took us to the yacht in a golf buggy. Arriving at the Marina and yacht we were greeted by the crew with a champagne and canapes reception on the rear deck. We felt privileged and special.

We settled in our staterooms and then returned to the lounge where the captain explained the route, the time scales, options and choices for stopping off at private luxury islands along the way. We cast off and headed away from the island, leaving behind what passes for a crowd on a private island. From there we head to a secluded anchorage and our first snorkel on the reef. The weekdays were filled with snorkeling, ski paddling, glass-bottom boat trips, jet skiing (both kept on board) and fishing as we move around the various reef waters taking advantage of sea conditions. While we were all experienced and familiar with water sport activities, it was our first-time scuba diving and reef fishing.

Both the captain and the first officer were PADI Divemasters and provided 1:1 instruction, while the chef supervised our catch and release, or keep for dinner, fish education. This included spearfishing for lobster as the basis for one evening meal and pairing each day's catch with wine choices from the 'cellar'. The 1:1 guest to crew ratio improved to 2:1 when two of our party had to return home, the yacht rushing to a floating helipad on the reef to connect them to an early commercial flight home.

After the evening meal, we would watch a movie or head to a 'closed' island for drinks at a bar normally exclusive to their paying guests. At the end of the week of cruising, we returned to the original island to re-join the homebound (other) 'tourists'.

Figure 3.2 begins to unpick the luxury charter experience by categorizing the events and opportunities made possible by plotting them as the 4Es of the experience economy (Pine & Gilmore,). Many of the situations and activities were similar to those documented in the inventory of wine-tourism experience offered by Quadri-Felitti and Fiore (2012). Their categorization formed the basis for plotting the rich points, that is the points from conversation or experience that require taking out for re-examination (Agar, 1980), that made up the luxury experience. While the model delineates four quadrants, as predicted by Quadri-Felitti and

Entertainment Tourists are engaged by the performance *Swimming, snorkelling, paddling, jet ski, fishing* *Sightseeing; reef, islands, private islands*	**Educational** Tourists enhance their knowledge or skills *Reef education* *SCUBA* *Catching food* *Selecting food and beverage for a menu with the chef* *Preparing food with the chef* *Reading charts and plotting courses*
Esthetic Tourists are enriched by sensual environments *Consuming the "reefscape"* *Enjoying the unique and luxurious lodging* *Fine wines and food*	**Escapist** Tourists become engrossed by participating in a different time and place *Escaping from other tourists with secluded moorings and small island hopping* *Nightclubs on private islands* *Helicopter transfers off the reef* *Catching own food*

Fig. 3.2 An inventory of the personal luxury yacht experience

Fiore (2012), the quadrants are permeable. Food, in particular, is central to the yacht experience and appears in all quadrants.

Treating the mention of food as a 'rich point' in the analysis (Agar, 1980), because of its unexpected appearance in every quadrant, suggests that there may be something uncommon about a luxury cruise on the Great Barrier Reef, aside from the staging of the experience on a mobile five-star resort. It is this ability for food to be the point at which the quadrants come together that delivers authenticity to the charter. Experiences surrounding food, i.e. combining *education* (about the marine life on the reef), *entertainment* (snorkeling with the fish, catching marine life for a meal later in the day), the *esthetics* of the yacht, reef, and the food presentation, and the escapism of not just the catch, but enjoying silver service dining on the site of the catch. In terms of authenticity, constructive and objective constructs are more easily achieved by any luxury reef vacation, however, the existential authenticity described here could only have been achieved with this exclusive luxury yacht charter reef experience.

Entertainment, i.e. 'when tourists are engaged by the performance' (Quadri-Felitti & Fiore, 2012: 8), while a passive participation experience, takes on an extra dimension when staged from a luxury yacht. Swimming and snorkeling were available at 'better times' than those offered as part of the scripted daily adventures by the reef cruising companies. If wind was predicted for midmorning (i.e. when the tour boats arrived at the reef), the captain ensured our (guest) snorkeling took place around breakfast. In addition, the luxury entertainment experience was exclusive. It included parts of the reef inaccessible to day-trippers, and visits to the quiet bars normally only available to 'residents-only' of the private islands.

Esthetics of experience, i.e. '*enrichment by sensual environments are magnified in the setting*' (Quadri-Felitti & Fiore, 2012: 8). Having exclusive and private access to small reefs, and the ability to move away if another craft moored in our line-of-sight allowed (complete) immersion in the 'reefscape' while enjoying the unique lodging and facilities of the luxury yacht which matches that of a five-star hotel. The rich sensual experience was magnified by the fine wine selection, the unique (fresh and local) meal ingredients delivered with silver service and at times convenient to guests.

The luxury yacht charter staff to guest ratio of 1:1 and 2:1 allowed extraordinary educational opportunities. Beginning as absorption, where guests '*enhance their knowledge or skills*' (Quadri-Felitti & Fiore, 2012: 8) we learned about the reef, about scuba diving, progressing to active participation by catching food. This continued through selecting ingredients and discussing the preparation of meals with the chef. Education also included learning to read charts and instruments alongside the captain as passages were planned through the maze of reefs, literally being immersed in the destination of the luxury yacht experiences.

Finally, the luxury yacht charter described offered total immersion in the Escapist realm of the luxury tourism experience. Characterized as 't*ourists become engrossed by participating in a different time and place*' (Quadri-Felitti & Fiore, 2012: 6), the escape began at the home airport and became more pronounced as the guests were separated from other island arrivals and were whisked away to the marina by the captain. From there the yacht allowed a literal escape from other tourists, even from another single yacht at an outer reef deep anchorage. Access to uninhabited islands, exclusive islands, helicopter transfers off the reef, the 'invisibility' of the yacht's crew unless bidden, contributed to the escapist experience of being alone far from the sight of land. This escapism was underscored by dinner planned around the fish caught daily by the guests.

Implications and Conclusion

Luxury, luxury tourism and the part of this market that is the luxury yacht charter market is growing. COVID-19 may provide further stimulus to this trend as consumers seek more exclusive and private holidays with their own small group where the itinerary and destination are controlled by them rather than the operator. Success in the luxury space, as in any market, is based on firms designing and delivering a competitive value proposition. Luxury is, however, an elusive concept. It is something which as individuals we somehow 'recognize' but also something that has not been so easily formulated, agreed upon and articulated. The volume of writing on luxury bears testimony to this.

This chapter has navigated the reader through some of the luxury literature and revealed some of its many facets. A characteristics approach suggests luxury exists in something by virtue of 'ingredients' that just have to be combined in a certain manner (for example, high-quality materials which may be exclusive). But luxury does not exist as something separate from the individual. It has an individual (subjective) dimension as well as a dimension that considers how one individual relates to another. There are private and public aspects to luxury. In this the role of symbolic meaning and signalling are important, i.e. the 'seeing' and 'being seen', and this has been a core element of understanding luxury. However, as Berthon et al. (2009) note, there is a shift away from the material and the symbolic to the experiential. Pine and Gilmore's 4Es framework (1998, 1999) provides a mechanism for considering the elements of the experience and, importantly, the role of the consumer can have in the co-creation of the experience (passive–active or absorptive–immersive).

In terms of the actual experiences, the luxury yacht charter provides a stage for consumers to express themselves in the design, co-production and consumption of the experience. A higher crew to guest ratio means that processes and itineraries can be bespoke, co-designed and co-produced with the guest more involved, more active and more immersed. However, this can only happen if producers (crew) facilitate this, providing opportunity where it is desired, or not as the case may be. It is a more fluid model of interaction between consumer (guest) and firm (crew). Staging memorable experiences is not, as per services marketing, about enacting set processes or doing things to consumers who are just receivers. Memorable experiences require different interactions and a different mode of operating.

In terms of the luxury yacht charter and the experience, there will be a role for object-related authenticity. For example, there is the yacht itself (for example, Italian furniture, a wine cellar, etc.). There is also the experienced crew. There are other elements too. For example, with a dining experience, object authenticity is founded on the ingredients of a meal. This may be reinforced through the process of meal dishes and in the savouring of them. There is also existential related authenticity too. For example, activities could include helping to crew a yacht, or diving or foraging for the ingredients for the dining experience or being educated or entertained in its preparation. Importantly, the possibilities for authenticity are likely to be greater on the luxury yacht charter, something that can itself be a source of what has so frequently been identified as important for luxury—*exclusivity*. In this part of the luxury tourism market, authentic experiences may serve to be a differentiating factor from other providers where the commodification of luxury (masstige) can make it lose its lustre.

A final point to close this section. Whilst the locus of luxury is shifting to experiences, symbolic value is still likely to be an important part of luxury. Symbolic value obtained from signalling to self and importantly signalling to others will continue to be a source of value. Consumers of a luxury yacht charter will likely want to be able to capture aspects of these experiences (e.g. a dive, dining or location) and share them to derive value. Of course, taking pictures and videos on mobile phones and using social media (e.g. Instagram) is one mechanism for facilitating this sharing and signalling. The consumer can capture in a photo or video any of the 4Es categories of experience. They can share their experiences and aspects of their felt authenticity. However, there is a potential dilemma even in this sharing of experiences; an 'Internet Dilemma' (Baker et al., 2018). This dilemma comes back to the important exclusive characteristic of luxury. The Internet (websites and social media) provides luxury firms and brands the opportunity for many potential benefits. For example, consumer reach and an ability to share rich information (breadth and depth beyond that of print media, etc.). Yet, with this technology challenges exist in establishing differentiation and uniqueness. The Internet's ubiquity (access by all), low barriers to entry or thresholds to high-quality design, presentation and interaction mean that establishing and maintaining a unique brand image is becoming increasingly difficult (Kapferer & Bastien, 2012). The signalling can undermine that pillar of luxury, the sense of exclusivity.

Finally, this chapter set out to provide an understanding of the key dimensions of luxury as a basis for understanding how a competitive value proposition for luxury yacht charters can be developed and delivered. This chapter detailed an important shift in focus in luxury thinking, from the material and symbolic to the experiential. Utilizing Pine and Gilmore's (1998, 1999) 4Es framework as a means of setting out how experiences can be categorized (and developed), and a partial ethnographic approach of a luxury yacht experience, this chapter highlights how each of the elements of the framework (entertainment, education, aesthetics, and escapist) come together to create a luxury experience. Important parts of this experience include co-design and authenticity. Indeed, these two elements are likely to be crucial in differentiating the luxury experience from other high-value experiences.

References

Adamczyk, A. (2015). How much does a superyacht really cost? *Forbes Life*. Retrieved May 1, 2021, from https://www.forbes.com/sites/aliciaadamczyk/2015/04/08/how-much-does-a-superyacht-really-cost/?sh=3100953536bc

Agar, M. (1980). *The professional stranger: An informal introduction to ethnography*. Academic Press.

Allied Market Research. (2019). *Luxury yacht market report*. Retrieved April 10, 2021, from https://www.alliedmarketresearch.com/luxury-yacht-market

Baker, J., Ashill, N., Amer, N., & Diab, E. (2018). The internet dilemma: An exploratory study of luxury firms' usage of internet-based technologies. *Journal of Retailing and Consumer Services, 41*, 37–47.

Berthon, P., Pitt, L., Parent, M., & Berthon, J. P. (2009). Aesthetics and ephemerality: Observing and preserving the luxury brand. *California Management Review, 52*(1), 45–66.

Boat International. (2021). *2001 Global order book.* Retrieved February 2020, from https://www.boatinternational.com/yacht-market-intelligence/luxury-yachts-on-order/2021-global-order-book

Bronner, F., & de Hoog, R. (2019). Comparing conspicuous consumption across different experiential products: Culture and leisure. *International Journal of Market Research, 61*(4), 430–446.

Cambridge Dictionary. (2021). Retrieved March 3, 2021, from https://dictionary.cambridge.org/dictionary/english/experience

CBI, Ministry of Foreign Affairs. (2018, November 21). *What are the opportunities for luxury tourism from Europe?* Retrieved January 10, 2021, from https://www.cbi.eu/market-information/tourism/luxury-tourism

Chen, C. F., & Chen, F. S. (2010). Experience quality, perceived value and satisfaction and behavioural intentions for heritage tourists. *Tourism Management, 31*(1), 29–35.

Chen, E. Y. I., Yeh, N.-C., & Wang, C. P. (2008). Conspicuous consumption: A preliminary report of scale development and validation. *Advances in Consumer Research, 35*, 686–687.

Correia, A., Kozak, M., & Reis, H. (2016). Conspicuous consumption of the elite: Social and self-congruity in tourism choices. *Journal of Travel Research, 55*(6), 738–750.

Credit Suisse. (2018). *Global wealth report 2018*. Retrieved March 25, 2021, from https://www.credit-suisse.com/about-us/en/reports-research/studies-publications.html

Danziger, P. (2006). *Shopping*. Kaplan Publishing.

D'Arpizio, C., Levato, F., Kamel, M.-A., & de Montgolfier, J. (2017, Fall-Winter). *Luxury goods worldwide market study*. Bain Report, Bain & Company.

Deloitte Italy. (2019). *Boating market monitor: Yachting market monitor in the luxury world*. Retrieved February 14, 2021, from https://www2.deloitte.com/content/dam/Deloitte/it/Documents/strategy/Boating%20Market%20Monitor%20in%20the%20luxury%20world_Deloitte%20Italy.pdf

Dhaliwal, A., Singh, D. P., & Paul, J. (2020). The consumer behavior of luxury goods: A review and research agenda. *Journal of Strategic Marketing*, 1–27. https://doi.org/10.1080/0965254X.2020.1758198

Dictionary.com. (2021). Retrieved March 25, 2021, from https://www.dictionary.com/browse/luxury

Di Domenico, M., & Miller, G. (2012). Farming and tourism enterprise: Experiential authenticity in the diversification of independent small-scale family farming. *Tourism Management, 33*(2), 285–294.

Froehlich, P. (2020, March 13). *Private yacht vacations are on the rise—And they may not be as expensive as you think, Travel and Leisure*. Retrieved March 24, 2020, from https://www.travelandleisure.com/travel-news/coronavirus-private-yacht

Gilmore, J. H., & Pine, B. J. (2007). *Authenticity: What consumers really want*. Harvard Business School Press.

Grbich, C. (2007). *Qualitative data analysis: An introduction*. Sage.

Gretzel, U., Fesenmaier, D. R., & O'Leary, J. T. (2006). The transformation of consumer behaviour. Tourism business frontiers: Consumers, products and industry. In D. Buhalis & C. Costa (Eds.), *Tourism business frontiers: Consumers, products and industry* (pp. 31–40). Routledge.

Gurzki, H., & Woisetschläger, D. M. (2017). Mapping the luxury research landscape: A bibliometric citation analysis. *Journal of Business Research, 77*, 147–166.

Hayes, D., & MacLeod, N. (2007). Packaging places: Designing heritage trails using an experience economy perspective to maximise visitor engagement. *Journal of Vacation Marketing, 13*(1), 45–58.

Hollebeek, L. D., Glynn, M. S., & Brodie, R. J. (2014). Consumer brand engagement in social media: Conceptualization, scale development and validation. *Journal of Interactive Marketing, 28*(2), 149–165.

Hosany, S., & Witham, M. (2010). Dimensions of cruisers' experiences, satisfaction, and intention to recommend. *Journal of Travel Research, 49*(3), 351–364.

Jadhav, A., & Singh, A. (2020). *Charter market: Global opportunity analysis and industry forecast, 2020–2027*. Allied Market Research.

Kapferer, J. N. (1998). Why are we seduced by luxury brands? *Journal of Brand Management, 6*(1), 44–49.

Kapferer, J. N., & Bastien, V. (2012). *The luxury strategy: Break the rules of marketing to build luxury brands* (2nd ed.). Kogan-Page.

Kapferer, J. N., & Valette-Florence, P. (2016). Beyond rarity: The paths of luxury desire. How luxury brands grow yet remain desirable. *Journal of Product & Brand Management, 25*(2), 120–133.

Ko, E., Costello, J. P., & Taylor, C. R. (2019). What is a luxury brand? A new definition and review of the literature. *Journal of Business Research, 99*, 405–413.

Mandhachitara, R., & Lockshin, L. (2004). Fast moving luxury goods: Positioning strategies for Scotch whisky in Thai department stores. *International Journal of Retail & Distribution Management, 32*(6), 312–319.

Manthiou, A., Lee, S. A., Tang, L. R., & Chiang, L. (2014). The experience economy approach to festival marketing: Vivid memory and attendee loyalty. *Journal of Services Marketing, 28*(1), 22–35.

Mehmetoglu, M., & Engen, M. (2011). Pine and Gilmore's concept of experience economy and its dimensions: An empirical examination in tourism. *Journal of Quality Assurance in Hospitality & Tourism, 12*(4), 237–325.

Mortelmans, D. (2005). Sign values in processes of distinction: The concept of luxury. *Semiotica, 157*, 497–520.

Oh, H., Fiore, A. M., & Jeoung, M. (2007). Measuring experience economy concepts: Tourism applications. *Journal of Travel Research, 46*, 119–132.

Park, E., Choi, B. K., & Lee, T. J. (2019). The role and dimensions of authenticity in heritage tourism. *Tourism Management, 74*, 99–109.

Phau, I., & Prendergast, G. (2000). Consuming luxury brands: The relevance of the 'rarity principle.' *Journal of Brand Management, 8*(2), 122–138.

Pianon, N., Abtan, O., & Bonelli, F. (2017). True global luxury consumer insight. In *altagamma.it* (4th ed.). The Boston Consulting Group, Fondazione Altagamma.

Pine, B. J., & Gilmore, J. H. (1998). Welcome to the experience economy. *Harvard Business Review, 76*(4), 97–105.

Pine, B. J., & Gilmore, J. H. (1999). *The experience economy*. Harvard Business School Press.

Quadri-Felitti, D., & Fiore, A. M. (2012). Experience economy constructs as a framework for understanding wine tourism. *Journal of Vacation Marketing, 18*(1), 3–15.
Skinner, B. F. (1953). *Science and human behavior*. MacMillan.
Song, H. J., Lee, C. K., Park, J. A., Hwang, Y. H., & Reisinger, Y. (2015). The influence of tourist experience on perceived value and satisfaction with temple stays: The experience economy theory. *Journal of Travel & Tourism Marketing, 32*(4), 401–415.
Spence, E. (2014). Unravelling the politics of super-rich mobility: A study of crew and guest on board luxury yachts. *Mobilities, 9*(3), 401–413.
Spence, E. (2016). Eye-spy wealth: Cultural capital and 'knowing luxury' in the identification of and engagement with the superrich. *Annals of Leisure Research, 19*(3), 314–328.
Trihas, N., Panagiotaki, G., & Kyriakaki, A. (2020). Insights into luxury tourism: The case of Elounda, Crete. *Journal of Tourism Leisure and Hospitality, 2*(2), 62–70.
Tsaur, S. H., Chiu, Y. T., & Wang, C. H. (2007). The visitors behavioral consequences of experiential marketing: An empirical study on Taipei zoo. *Journal of Travel & Tourism Marketing, 21*(1), 47–64.
Turunen, L. L. M., & Laaksonen, P. (2011). Diffusing the boundaries between luxury and Counterfeits. *Journal of Brand Management, 20*(6), 468–474.
Tynan, C., McKechnie, S., & Chhuon, C. (2010). Co-creating value for luxury brands. *Journal of Business Research, 63*(11), 1156–1163.
Van Boven, L., & Gilovich, T. (2003). To do or to have? That is the question. *Journal of Personality and Social Psychology, 85*(6), 1193–1202.
Veblen, T. (1899). *The theory of the leisure class*. MacMillan.
Vigneron, F., & Johnson, L. W. (1999). A review and a conceptual framework of prestige-seeking consumer behavior. *Academy of Marketing Science Review, 1*(1), 1–15.
Vo-Thanh, T., & Kirova, V. (2018). Wine tourism experience: A Netnography study. *Journal of Business Research, 83*, 30–37.
Wang, N. (1999). Rethinking authenticity in tourism experience. *Annals of Tourism Research, 26*(2), 349–370.
Wang, Y., Feng, Y., & Feng, B. (2013). The study on the significance of difference between demographics and tourist experiences in Macau Casino hotels. In *Proceedings of 2013 International Symposium—International Marketing Science and Information Technology*.
Yang, W., & Mattila, A. S. (2017). The impact of status seeking on consumers' word of mouth and product preference: A comparison between luxury

hospitality services and luxury goods. *Journal of Hospitality & Tourism Research, 41*(1), 3–22.

Yeoman, I. (2011). The changing behaviours of luxury consumption. *Journal of Revenue and Pricing Management, 10*(1), 47–50.

Yeoman, I. (2012). *2050-Tomorrow's tourism* (Vol. 55). Channel View Publications.

Yeoman, I., & McMahon-Beattie, U. (2006). Luxury markets and premium pricing. *Journal of Revenue and Pricing Management, 4*(4), 319–328.

Yeoman, I., & McMahon-Beattie, U. (2011). The changing meaning of luxury. In I. Yeoman, & U. McMahon-Beattie (Eds.), *Revenue management* (pp. 72–85). Palgrave Macmillan. https://doi.org/10.1057/9780230294776_7

4

Luxury Yachting versus Luxury Cruising: Differences, Similarities and Customer Choices

Karen Worrall and Tatiana Gladkikh

Introduction

Luxury yachting tourism used to be the domain of only the super-rich who could afford their own vessel worth tens of millions of Euros (Spence, 2014). The luxury yacht industry has more than doubled in the last ten years (Solent University, 2021), and in particular, demand has grown since the COVID-19 pandemic engulfed the world and brought the global tourism industry to its knees in March 2020 (Ito et al., 2020). It is becoming more on the radar as a segment of Special Interest Tourism (SIT). Nowadays, more advertising is being displayed about luxury yachting holidays (Ingram, 2021), highlighting its increase in popularity as an alternative to luxury cruise holidays with a higher

K. Worrall (✉)
University of the Highlands and Islands, Inverness, UK

T. Gladkikh
The University of the Highlands and Islands, Perth, UK

T. Gladkikh et al. (eds.), *Luxury Yachting*,
https://doi.org/10.1007/978-3-030-86406-4_4

number of guests sharing the same holiday vessel (Mitruka & Wheeler, 2008).

The aim of this chapter is to investigate the major aspects of these two interlinked sectors of seagoing tourism and to evaluate the amenities, services and activities which make guests choose one or the other. Major concepts that will be investigated in this chapter are luxury and class; exclusivity and access; privacy and different categories of service and experience; small-ship cruise experience and environmental impact. The limited number of publications on the luxury yachting industry compared to that on the cruise industry is a challenge faced in writing this chapter. However, the production of this book aims at addressing the gap.

Our chapter is informed by real-life observations and experiences of one of the authors, who has worked in the luxury cruise industry since 2005. This experience of working on six different luxury cruise lines and one small-ship luxury cruise line offers insight into the issues discussed in this chapter, giving it a more personalized angle. There are also some useful comments from two other practitioners who worked in the luxury yacht sector of the industry as a chef on a 24-metre luxury yacht, and a stewardess on 35-metre, 45-metre and 54-metre yachts. Specific names of cruise lines, ships and yachts have been omitted for confidentiality reasons, except when discussing general trends.

Utilizing personal experience and expertise gained from working in the industry is a useful approach in research. It brings a practitioner's perspective of seagoing tourism into investigation and helps to advance our understanding of it through practice-based theory building. It is a helpful way to generate new knowledge. It assists in bridging the gap between rigour and relevance of academic–practitioner relationships, as highlighted by Bartunek and Rynes (2014). Practical first-hand knowledge of a subject is not always utilized in academic work as preference is usually given to using existing theories. In subjects where practical experience gives insights into otherwise hidden worlds, such as luxury yachting, it can be very useful as it links practical knowledge with theory which provides a more complete picture and offers solutions to solve problems, such as through formulation of policy and management actions (Acharya et al., 2020). Bringing practice and theoretical developments together also helps to avoid stereotypes of researchers being too removed from the reality of their study and practitioners being too close to it to see the broader picture, thus bringing the two viewpoints together makes a more effective study (Warwick University, 2021).

In an attempt to understand luxury yachting and luxury cruising better, it is worth mentioning the small-ship cruise experience, which functions in some ways as a hybrid of a large luxury yacht and small luxury cruise ship. It is an interesting section of the industry that fits between the two other types (luxury yachts and luxury cruises). It involves very small cruise ships (usually with between 300 and 600 guests) which are sometimes seen to have the best of both worlds. They are more luxurious than cruise ships generally, have excellent culinary features, have more choices in entertainment than yachts, very high levels of attentive customer service and are more private than cruises but have more social options than yachts. This will be factored into our discussion.

Sea-Tourism Industry and its Segments

The luxury cruise industry is immense. It grossed $154 billion to the worldwide economy in 2019 pre-COVID-19 (CLIA, 2021), its revenue is generated from 323 cruise ships sailing worldwide (Cruise Market Watch, 2021a) belonging to more than 70 ocean and river cruises lines (Lipcon et al., 2021). The industry is the fastest growing sector of tourism, with demand estimated to grow at 7% per annum over the past decade (Syriopoulos et al., 2020) with more than 30 million guests cruising in 2019 (Wondirad, 2019).

Sizes of cruise ships range greatly. There is some debate over which is the smallest regular cruise ship. One example could be Regent Seven Seas Navigator which has 490 guests, weighs 28,803 gross tonne (GT) and is 171 metres long (Regent Cruises, 2020).

Currently the world's largest cruise ship is Royal Caribbean's Symphony of the Seas, which carries 5518 guests at double cabin capacity with 6680 at full capacity using all pull down bunk beds designed for children. It is 362-metres long and weighs 22,800 GT (RCCL, 2021).

The luxury yachting industry was worth $64.1 billion in 2020 (Global Industry Analysts, 2021) and is expected to reach more than $83.2 billion by 2025 (ReportLinker, 2020) with a growth rate of 4.1% between 2020 and 2027 (Global Industry Analysts, 2021). The most popular size of luxury yachts is between 20 and 50 metres in length, which accounts for more than half of all luxury yachts (Grand View Research, 2021b). There are currently more than 5000 yachts longer than 24 metres in service, with more than 30 main companies producing them (Solent Uyniversity, 2021).

An in-between variant that lies in size and service between luxury yachts and luxury cruise ships is small-ship cruising. This category can carry from 62 guests at a size of 82-metres long, at a gross tonnage of 3300 GT, like the Crystal Esprit (ROL Cruise, 2021) to Windstar, mainly with 250–350 guests, around 10,000 tonnes (Windstar, 2020) to Seabourn having usually 450 guests, with ships of 198-metres length and weighing in at 32,346 GT (Seabourn, 2021b).

According to the Florida-Caribbean Cruise Association (FCCA), the top three elements that attracted people to book a luxury cruise in 2006 involved being pampered, visiting multiple destinations in one trip and fine dining (FCCA, 2006). The top three elements that attract people to book a luxury yachting holiday involve exclusivity and privacy, access to places larger vessels aren't permitted to visit (Levius, 2020) and high levels of personalized service (CBI, 2018).

COVID-19 and its Effects on the Sea-Tourism Industries

COVID-19 pandemic brought worldwide tourism to its knees (Uğur & Akbıyık, 2020). COVID cases on cruise ships were often exaggerated in the media causing panic and fear. Unfortunately, any studies focusing on what was done right and lessons to learn for potential future infectious disease outbreaks only came later (Brewster et al., 2020). As the pandemic evolved, cruise ships were still vilified and the industry has still not fully restarted or recovered, with a large number of ships still being on warm lay-up (Hancock, 2020).

As restrictions began to ease somewhat in summer 2021 and some travel has been allowed for a time at least in some countries, luxury yachting has been able to restart (Steele, 2021).

Why People Cruise and Why People Yacht

The lure of the sea has been something that has called to humans for at least 4000 years, since the voyagers of Taiwan moved South and East to Polynesia and beyond to find new islands (Soares et al., 2011). What began as a way to find better food sources, more hospitable climates and easier to live in areas—as well as adventure—is now something thousands of people spend their hard-earned pay on each year to get away.

While no specific figures could be found on the number of people taking trips on luxury yachts annually, it is estimated that 830 new yachts were constructed in 2019 (Montigneaux, 2019), therefore it can be assumed that annual numbers are at least in the tens of thousands. The question arises: why do some people choose cruising and why do some choose yachting?

One of the main differences between the two industries is size. With ships, size matters. The size of the vessel affects the size of guest and crew count, and amount, calibre and exclusivity of amenities available on board. Luxury yachts are much smaller in size and have a higher price tag following the economy of scale (Ang & Lin, 2001). They have more expensive features per head, including décor, ratio of staff to guests and food. This means that they are very high-end in how they look with furnished details, have higher and more personalized levels of service due to generally a 1:1 guest to crew ratio (Spence, 2014). According to the industry professionals, some luxury yachts have a ratio of 1:2 which can go to as low as 1:0.5.

Luxury cruise liners are much bigger in size; therefore, they are more cost-effective overall. Ship décor is usually designed to be elegant but also hard-wearing to ensure durability due to extensive use by numerous passengers. The guest to crew ratio is between 1:1.5 (with Crystal at the top end) down to some main cruise lines measuring 1:0.3. However, on average luxury cruise lines generally provide the 1:1 to 1:0.4 ratio, which translates to one crew member for approximately every one or two guests (Cruisewatch, 2021).

Bearing these main points in mind, luxury yachters generally spend more money on their trip per person than their cruise counterparts, as they are offered more exclusivity, privacy, access and higher levels of service and calibre of food. While these aspects are generally of higher quality on luxury yachts, cruise ships offer other appealing elements which yachts can not provide due to their size. One of them is entertainment on board. The larger the vessel, the more choice there can be available on it for its guests. Major other desirable aspects of choice include food and social options. These aspects will be discussed in more detail further in this chapter.

Luxury and Class

Luxury can be difficult to define as it is something that can be deeply personal. The concept of 'my luxury' can imply a personal decision of how someone chooses to spend their time or money. For someone with a lot of money and little time, luxury can be having the time to spend with loved ones (Kapferer & Bastien, 2012). For others, it can be the feeling of being pampered and looked after, without having to do the work, having staff take care of everything for you (Spence, 2014). For others, luxury on holiday can mean having excellent dining and entertainment facilities (Bakker, 2005).

Kapferer and Bastien (2012) highlight that luxury isn't something that just 'is' or an object that can be analysed. It is the output of social dynamics of a society which differs in different societies (Bakker, 2005). Kapferer and Bastien's functional perspective on luxury segmentation posits that luxury can be about personal experience for oneself, using it for personal expression and an art of living, or to signal to others their power through their sense of fashion, or 'membership' to reassure their class or taste (Kapferer & Bastien, 2012).

Class plays a major factor in most industries. Kraus et al. (2017) suggest that there are four main types of class—lower, working, middle and upper. Linking class considerations to people's financial situation and affordability of holidays, this generally means that the lower class are always struggling; working class are able to pay the bills but probably don't have too much extra for unnecessary things such as holidays; middle class don't worry about money as they are comfortable and able to afford holidays; upper class are wealthy enough to not have to consider costs in any decision-making. Kraus et al. (2017) posit that class signals help individuals to feel more connected to those they perceive in the same class as them, and give class signals, which are strong in determining the likelihood of formation of relationships and offering access to influential networks. This is an important factor in luxury yachting, as fellow owners or charterers feel that they belong to the same 'club' (Kraus et al., 2017).

The above discussion of the meanings of luxury are applicable to the sea tourism and can be relevant to both a luxury yacht and a cruise trip.

Signalling Wealth

As argued earlier, luxury yachting is generally more expensive than luxury cruising. The exclusivity, access, privacy and (typically) higher level of service come at a price, therefore luxury yachting trips affirm one's status. However, high status can also be affirmed and signalled from cruising on a luxury cruise ship. Cruise ships have various grades of staterooms to choose from, with more expensive and larger ones offering more amenities and a more 'luxurious' experience, which can be experienced through extras such as more attentive cabin stewards and in-stateroom benefits, priority access to take ship's tenders ashore and complimentary meals at speciality restaurants.

Most cruise lines offer a special concierge-attended lounge. Personal observations in the industry show that such lounges offer complimentary amenities only available to suite guests. That includes drinks such as barista-grade coffees, sparkling and still bottled water, soft drinks, wine, liquor; special snacks such as croissants and other freshly baked pastries in the morning, hand-cut smoked salmon sandwiches and mini-quiches in the afternoon and chef's special macaroons, available nowhere else on board; up-to-date high-level magazines, newspapers and books such as the daily most high-level newspapers, National Geographic, The New Yorker, along with up-to-date travel books and elegant coffee table style art books. The staff in the concierge lounge are always of the highest levels of training and equipped to cater to any guest requests. For example, they would help to arrange exclusive appointments for guests in the lounge with other relevant staff members. This might happen in a situation when the guests wish the onboard cruise consultant to visit them personally to book their next trip. Such a lounge provides an area where guests can meet other high-paying guests and form social connections.

Personal experience of working on cruise ships has allowed to observe that guests often love to signal their 'luxury status' to anyone who will listen to them. Many will tell the crew how many cruises they have been on. Chaleila (2019) points out that this type of signalling usually comes from the 'nouveau riches' who are trying to emulate the cruise ship 'aristocracy' and who can resemble the characterized version of Molly Brown in the movie Titanic. Guests who are truly very wealthy usually don't feel the need for this showmanship, which seems to be human nature across all industries (Chabrak et al., 2016).

Demographic

Cruise lines have had a reputation of being a place where older generations vacation. This tendency is changing. Currently, various cruise lines target different demographics in relation to age. In the author's experience from working for cruise companies, supported by evidence from colleagues, generally, older guests with median age of 65+ enjoy sailing on Holland America Line, Cunard and Silverseas. Couples mainly from the 40–65 bracket enjoy Princess cruises and Viking. Royal Caribbean Cruise Lines and Disney are favoured with families as they cater very well to children. Carnival does not generally cater to the luxury demographic but offers young adults a 'party cruise' experience.

Luxury travellers are getting younger. As Baby Boomers are making way for 'Generation X', 40-year-olds are becoming the new 50-year-olds in terms of disposable income. Many affluent luxury travellers are now in the 30–40 age band (Bakker, 2005). This can vary from country to country (Bakker, 2005), but the overall trend is that luxury cruisers or yachters cannot be stereotyped into a particular age or nationality bracket. The only stipulation for admission is affordability.

Exclusivity and Access

Two of the aspects of choosing a luxury yacht holiday over a luxury cruise holiday involve exclusivity and access. Fewer people can be accommodated on a luxury yacht, therefore this restriction immediately stipulates being part of an elite club. Cruise ships are much larger and accommodate significantly more people. The fares for cruising vary depending on the category of stateroom (cabin). Much of the holiday experience outwith the stateroom is the same, regardless of the category of cabin, as most options for dining, entertainment and usage of the ship's main lounges (except for those reserved for suite guests) are available to all.

In contrast, the experience on a luxury yacht is very exclusive. Many yachts offer butler-level service, with more one-to-one attention and personalized service, compared to offering a list of services for all to choose from, as generally happens on cruise ships. Yachts tailor individual service specific to their guests, raising the exclusivity factor. From the first-hand experience of yachting practitioners, the crew are expected to meet the wishes of their guests on command, including preparing meals in the middle of the night or arranging parties at last-minute notice. Cleaning standards involve cleaning surfaces with toothbrushes and toothpicks, and vacuuming magazines. The crew have to be very experienced to be hired, they usually undertake extra courses in sea survival and silver service to become a steward(ess). To be offered employment on a yacht usually requires the applicant to be introduced into the world by a contact.

Luxury cruise ships can offer several ways of experiencing exclusivity within the larger context of the ship. For example, the guests staying in suites may have access to private attended cabanas (tent-like lounge areas which can generally accommodate up to four people) in a specially designated part of the ship which is unavailable to regular guests. However, these areas can be accessed by any guest who is prepared to pay the per-cruise fee, which is usually c.$300 per person per week.

Many cruise lines offer extra benefits that come automatically with the suites, but also make some of these available to book for an extra charge to any guest who wishes to experience them. This means that extra exclusive access can be arranged at an extra charge. These include such services as specialist restaurants which offer a higher level of service which can be available as a paid extra of usually around $15–40, and access to the ship's spa's thermal suite, which would typically cost around $200 per couple for a week-long cruise. For some guests staying in less expensive inside staterooms these added luxuries afford the experience of exclusivity at a much lower cost than booking a suite (Radon, 2012).

Another element of access and exclusivity to consider is vessels' docking location. Cruise ships are larger and therefore physically unable to dock in every port. Owing to their smaller size, yachts generally are able to dock in very small ports inaccessible to ships. There is rarely a need to tender in to a port, unless the port is full and tendering in allows access on a busy day. This also allows to save money on berthing fees.

The experiences available ashore are generally the same for guests on yachts and cruises. Shore excursions in any port are limited to what the area has to offer. Luxury yachts might have access to more exclusive tours by paying a premium but cruise ships also offer different levels of tours. For example, one can choose a tour of Florence on a bus with 40 other guests or to a private car journey with a driver guide for 2–4 people. Both types of holiday would have access to negotiating their own independent tours with local companies.

Privacy

Privacy is something that luxury yachts can offer with more certainty than luxury cruises, which are larger in size and accommodate more people. Traditionally, luxury yachts offer privacy from other vacationers and the general public, which is highly valued by people in the spotlight. There is also evidence that rich tourists like to vacation with other rich tourists (Batabyal & Yoo, 2020), which can help explain the popularity of chartering luxury yachts, as well as the small-ship cruise options. A social aspect of wanting to be around one's 'tribe' of the fellow elite could be at play here (Kapferer & Bastien, 2012).

Part 2: Experiences on Offer

Part two of this chapter offers some practical observations of experiences on both segments of sea tourism and discusses the appeal of both. Adding these observations to the discussions above allows conclusions to be reached about deeper meanings of their appeal.

Social Aspects, Entertainment and Enrichment

Becoming acquainted with new people plays an important part in choosing a holiday on a cruise ship versus a luxury yacht. Yachts are smaller and can accommodate fewer people. The main options are to own one's own to enjoy privacy with own family and friends or to charter someone else's yacht for comparable purposes. Meal times are often a shared event on a yacht. There are limited public areas to relax on board, hence people tend to meet in shared spaces.

Cruise ships are bigger and accommodate significantly more people. They are very convivial and amiable places. This makes cruise ships one of the most popular choices for vacations for solo travellers. The environment facilitates meeting new people on board and forging dinner companions in a ship's sociable atmosphere. The quality of these interactions also plays an important part in the role of the traveller's experience (Huang & Hsu, 2009).

A larger number of sociable entertainment options available on cruise ships means that the likelihood of meeting people with mutual interests increases too. For example, events such as ballroom dancing, line dancing and trivia are usually very popular and many guests make new friends attending these events (Huang & Hsu, 2009). This is wonderful for those seeking gregarious interaction, but not for people who want their holiday to be a holiday not only away from work but also away from people.

For this reason, the desire to be away from other travellers and have a genuinely exclusive retreat is one of the biggest draws to yachting, as observed through practice.

Entertainment on cruise ships and luxury yachts is very different. Reflecting the social aspect, luxury yachts are much smaller and offer limited variety of live entertainment on board. Cruise ships usually have a full-sized theatre with a main show every night, with a different act performing two shows every night. This includes international high-level talent acts from singers to comedians, instrumentalists, illusionists and more. Information about available entertainment options can be found on the websites of all major luxury cruise lines, including Holland America Line (Holland America Line, 2021), Princess Cruises (Princess Cruises, 2021), Royal Caribbean International (Royal Caribbean, 2021), P & O Cruises (P&O, 2021), and Seabourn (Seabourn, 2021a). There are usually several cocktail lounges with different genres of live music on offer and lounges with interactive events including game shows. Similar experience can be found on small-ship luxury cruises.

Luxury yachts usually do not offer traditional entertainment. 'Entertainment' on luxury yachts is typically what would be categorized as 'activities' on cruise ships. This includes utilizing the yacht's own water sports equipment, such as jet skis, kayaks, paddle boards and snorkelling equipment, which the crew would set up for the guests. Other 'entertainment' would include an option to access small private islands and bays by yacht that cruise ships wouldn't often have access to.

Enrichment is a growing concept on luxury cruise ships, offered by companies that value learning as a contextualized part of the guest experience. Small-ship luxury cruises and exploration vessels use this in their standardization of entertainment enhancement. Travellers who aspire to demonstrate their intellect and eco-responsible side, choose to travel with purpose (Lew, 2018) to learn about the places visited, their local culture, food, history and wildlife. This segment of tourism is growing. More cruise ships and small-ship luxury cruises hire lecturers and presenters to give talks on in their areas. For example, naturalists are hired in Alaska to discuss the wildlife which is possible to see from the ship (Manley et al., 2017) and Polynesian ambassadors are contracted on itineraries in Hawaii and the South Pacific to teach hula dancing, ukulele classes or make fresh orchid leis. Academics are hired on multiple itineraries to teach about historical and cultural aspects of the destinations, including experts in the Moai history in Easter Island and marine biologists for Australian cruises (Travel Weekly, 2021).

Culinary Experience

The culinary experience is another principal part of the allure for sea tourists. The calibre of cooking varies between luxury cruise ships and luxury yachts and is usually determined by the budget for food, number of people on the vessel and the variety of cuisine available. Cruise ships generally have a lower monetary allocation for food, with their daily meal costs per capita averaging at around $12–13 (Cruise Market Watch, 2021b). Planning meals on cruise ships is a colossal task, since it involves meticulous planning to calculate how much food will be required. From personal observations in the industry, entire departments are dedicated to the logistics and supply chain management estimating where to purchase food, how much food will be needed and to ensure suppliers are available in strategic ports for longer voyages. Generally, cruise ships can carry around 10 days' worth of food supply to accommodate all guests and crew. This means that there is plenty of choice, but the choice is predetermined for the whole trip before it begins.

A notable benefit of food on cruise ships is the variety available. Cruise ships offer multiple restaurants on board with a wide variety of cuisines available. This usually includes at least a main dining room for dinner (breakfast and lunch on request), a general buffet restaurant with a menu theme that changes daily for lunch and dinner options (usually offering the same breakfast options daily), two or three casual dining spots such as a burger joint, taco bar, salad bar or pizza place and optional extra speciality restaurants which have nominal charges, such as a steak house, speciality Italian and Asian restaurants. Thus, travellers who enjoy such variety would enjoy the cruise culinary experience. The main dining room also offers guests the opportunity to socialize with different people every night if they wish to make new friends. It also allows getting away from the guests they do not wish to meet.

Luxury yachts generally have one dining area for the guests, with a more limited choice of food options. Practitioner observations confirm that some yacht owners enjoy having food purchased by chefs in the local area and meals created from locally sourced produce using their chefs' creativity, culinary skills and knowledge. This involves a somewhat higher daily spending per head on meals. Having the ability to curate their own menu locally is one of the benefits of luxury yachting. It means the meals can be tailored to meet the demands of the guests. It also means that part of their experience on the yacht can be brought into mealtimes, with guests able to fish in the sea with their catch being cooked for them straight away. Chefs who work on luxury yachts have to be highly qualified. They often use locally sourced, organic, high-quality ingredients that aren't available at scale. Generally, chefs prepare a menu for the week in advance and the yacht owners or charterers would approve or amend it. Provisions are normally purchased along the way if it was a longer trip or the owners wanted to try something in particular from where they docked.

Both options offer a great culinary experience, but on yachts, there is more ability for the guests to curate their own menu.

Environmental Impact

There is more societal pressure now than ever before to 'do no harm' in business, including in tourism. Overtourism is a global issue, and COVID-19 has highlighted the importance to preserve our precious world (Ioannides & Gyimóthy, 2020). Stricter legislation has been passed in the past decade for the cruise industry after 'mistakes' have been made, which forced the industry to adhere to stricter rules in regard to pollution of waters (Copeland, 2011).

Luxury yachting delivers less negative impact, and their guests tend to be more sustainably and environmentally demanding (Zolfani et al., 2015). Greener yacht features and products which are the preference of eco-conscious yacht owners include new yacht designs being a fifth of the weight of older models to cut fuel consumption, yachts fitted with solar panels that produce LED power and electric illumination and using environmentally friendly cleaning products (Mallet, 2010). The higher price tag of luxury yachting allows appropriate changes to be implemented, for example through installing such features as hybrid propulsion systems to reduce underwater noise pollution and fuel usage by 30%, waste heat recovery systems and integrated batteries for optimal performance (Boat International, 2021). Some yacht-making companies are dedicated to reducing and negating negative environmental impacts through Blue Card rules, CE Certifications and Blue Flag policies (Sevinc & Güzel, 2018).

Small-Ship Cruising: the Best of Both Worlds?

An important aspect of the luxury cruising and yachting industry to consider is small-ship luxury cruising. These are high-end six and seven-star ultra-luxury cruise lines, whose size and features lie in between luxury cruise ships and luxury yachts. This type of small-ship cruising is offered by such companies as Seabourn (2021b), Regent Seven Seas (2021) and Silverseas (2021). These can be seen as the best of both worlds option. They have around 300–600 guest capacity, with only a couple of ships reaching the 750 guest mark in this category, as opposed to cruise ships usually holding more than 1000 guests and luxury yachts usually having fewer than fifty.

The demographic of guests on small-ship luxury cruises is varied. The guests are generally somewhat younger than on many luxury cruise ships. These small ships are more petite and sleek, and therefore able to visit more adventurous locales, which attracts younger clients. Small-ship cruising is expanding as an industry as observed by Conde Naste Traveler and this segment will continue to grow in the post-COVID-19 world (Eskins, 2021).

Discussion and Predictions for the Industry

The observations and discussions of luxury yachting and luxury cruising presented above and supported by an account of the small-ship luxury cruise option, allow us to return to Kraus et al.'s class model (2017) and propose that, while at the moment luxury yachting is only affordable to the upper class, the growing levels of monetary wealth and the increasing desire for privacy and experience gradually create a new 'middle-class' option for luxury sea travel, which will arise through revenue management of price optimization and finding the optimal maximum price customers are willing to pay. This is similar to Ayvaz-Cavdaroglu et al.'s (2019) method of partial substitutability among products, which suggests that to create a similar luxury, or pseudo-luxury experience at a lower price point, the product can be accomplished by partially substituting some of the luxurious aspects of it. For luxury yachts, or some aspects of the experience on them, this could include using cheaper materials to build yachts and choosing less extravagant and cheaper materials for the yacht's décor, creating food menus with a lower culinary budget, sailing to ports which are cheaper to dock or anchor in and being less stringent with calibre of crew, thus saving money on wages.

This won't take anything away from the true upper-class luxury yachting but would add more of an achievable middle-class layer to it. This in itself is a paradox, as luxury is associated with being expensive, but the industry grows through being accessible (Kapferer et al., 2014), so if it becomes affordable to the 'middle class', can it still be marketed as luxury? It can be argued that ideal marketing solutions to this paradox involve contextualizing the creation of mass prestige (Moreau et al., 2020) and operationalizing masstige marketing techniques, following the example of many luxury brands (Kapferer & Bastien, 2017), giving the impression of exclusivity in yachting that becomes affordable for more people, using goods produced by luxury brands intended for the mass market (Collins Dictionary, 2021). Giorgio Armani auspiciously conceptualized this when he conceived his sub-brand Emporio Armani, then Armani Exchange to make the brand more accessible, by being more affordable, and devising disparate themes to the sub-brands, with

Giorgio Armani being the original line designed by the man himself, Emporio Armani having a formal theme, while Armani Exchange has a more everyday feel (Fearnley, 2021).

Here we introduce the term 'masstigification' of the luxury yacht industry, which has emerged as a result of the evaluation of the likelihood that mass prestige and masstige branding will be utilized in the marketing of luxury yachting trips. The term refers to the prediction discussed earlier that luxury yachting will become somewhat more accessible due to higher demand on charter yachts as a result of COVID-19. This term is expected to gain more relevance as luxury yachting companies will continue to broaden their customer base by creating slightly lower calibre vessels with comparable amenities, staff and food. It is anticipated that luxury yacht companies will—or should—capitalize on the success of masstigification that other luxury brands have successfully executed, such as Armani.

It is expected that luxury yachting will continue to grow, becoming a better known and more actively used segment within Special Interest Tourism. People feel safer booking a holiday where they will be around fewer people, and the idea of a small luxury boat taking you around secluded islands is very appealing. According to the Grand View Research's forecast (Grand View Research, 2021a), luxury yachting will continue to increase in demand among those who can afford it due to the new more crowd-wary consciousness of the emerging post-COVID-19 world. If this projection is true for luxury yachting and more people holiday on them, then the exclusivity of the luxury yacht sector will to a certain degree be exposed to erosion, as guests share their stories and images of their experiences on social media, and information about luxury yachting becomes more easily accessible. Nevertheless, a prominent space for privately owned luxury yachts operating exclusively for their owners will continue to offer their private holiday experiences. Privacy will continue to be the main attraction for this top-end segment with the upper class.

Observations suggest that demand for small-ship cruises will continue to grow in popularity (Williamson, 2021) as they are able to offer experiences in-between luxury yachting and luxury cruising, providing their guests with the benefits of the 'best of both worlds'. With COVID-19 in mind, having a smaller guest count on a small ship possibly makes cautious travellers feel more at ease than on a larger ship at the moment, even though they still offer more options for entertainment, socializing and culinary experiences than luxury yachts.

The customer is king, and if demand for the luxury yachting sector and/or small-ship cruises continues to rise, major luxury yachting companies, such as Royal Van Lent's Feadship and Princess Yachts, both owned by the world's largest luxury conglomerate LVMH (Moët Hennessy Louis Vuitton) (LVMH, 2021) could do well to join the masstige party, and create their own brand (Moreau et al., 2020). This will allow them to bring their brands of luxury from solely the upper-class domain into the middle-class sphere, thus increasing their market share overall. Guests who opt for the theoretical 'middle-class' luxury yacht options now, may select the 'upper-class' highest level experience later, thus adding the extension of a lower level now will not negate aspirations for the highest level, but leaves guests with something to aspire for.

It is expected that luxury cruising will recover from the COVID-19 pandemic, with cruise bookings for 2022 already higher than they were for 2019 (Rizzo, 2021). However, due to multiple delays of cruising resuming throughout 2020 and 2021, it is likely that full confidence in cruising will not return until sometime in 2022 when a much larger proportion of the population has been vaccinated globally making sea travel a more enjoyable and relaxing experience.

Conclusion

This chapter has discussed the differences between luxury yachting and luxury cruising, taking into account small-ship cruising as a cross between the two sectors of sea tourism. The concepts of luxury and class informed the evaluation of preferences of guests choosing yachting versus cruising. It was highlighted that privacy, exclusivity, access and high-level service and high-calibre culinary experiences were among the most valued aspects for yachting guests, while having more options for social interactions, entertainment, enrichment and culinary experiences were major factors in choosing luxury cruise vacations. It is worth noting that class and wealth play a part in this, as luxury yachting is only affordable to a limited number of high-net individuals, while cruising has different levels of pricing available, ranging from higher to lower level staterooms within the same ship. This makes cruising more affordable to a wider range of guests.

Owing to the current global condition dominated by COVID-19 which has influenced more cautious attitudes of travellers, luxury yachting and small-ship cruising are predicted to continue to grow in popularity. As for luxury cruising, it may take some time to establish a new form of 'normal', although high numbers of bookings for 2022 show promising signs of recovery for the industry. Considerations were also made regarding the luxury yachting industry opening up to 'masstigification', thus creating 'middle-class' yachts to satisfy the demand. What remains certain is that the appeal of sea travel and holidaying will continue attracting both the loyal fan base of cruisers and fans of yachting, which means that there is space for luxury yachting, small-ship cruising and luxury cruising.

Acknowledgements Our special thanks go to executive chef, Drake 'Chef Hollywood' Robertson (https://chculinarycreations.com/) and former stewardess Abbie Cripps, who provided their valuable insight into working on a 24-metre luxury yacht and on a 54/45/35-metre yacht respectively. Thank you.

References

Acharya, K. P., Thapa, R. K., Kuwar, K. J., Thapalia, B. P., & Paudel, P. K. (2020). Policy and management actions that resulted in curbing rhinoceros poaching. *Journal of Applied Ecology, 57*(8), 1452–1458. Retrieved from https://doi.org/10.1111/1365-2664.13692

Ang, J. S., & Lin, J. W. (2001). A fundamental approach to estimating economies of scale and scope of financial products: The case of mutual funds. *Review of Quantitative Finance and Accounting, 16*(3), 205–222. Retrieved from https://doi.org/10.1023/A:1011232319447

Ayvaz-Cavdaroglu, N., Gauri, D. K., & Webster, S. (2019). Empirical evidence of revenue management in the cruise line industry. *Journal of Travel Research, 58*(1), 104–120. Retrieved from https://doi.org/10.1177/0047287517737178

Bakker, M. (2005). *Luxury and tailor-made holidays*. Retrieved from https://doi.org/10.13140/RG.2.1.2358.6968

Bartunek, J. M., & Rynes, S. L. (2014). Academics and practitioners are alike and unlike: The paradoxes of academic–practitioner relationships. *Journal of Management, 40*(5), 1181–1201. Retrieved from https://doi.org/10.1177/0149206314529160

Batabyal, A. A., & Yoo, S. J. (2020). A theoretical analysis of preference matching by tourists and destination choice. *Asia-Pacific Journal of Regional Science, 4*(3), 809–820. Retrieved from https://doi.org/10.1007/s41685-020-00160-w

Boat International. (2021). *Sustainable superyachts: The world's most eco-friendly yachts*. Retrieved from https://www.boatinternational.com/yachts/yacht-design/eco-friendly-superyachts--25627

Brewster, R. K., Sundermann, A., & Boles, C. (2020). Lessons learned for COVID-19 in the cruise ship industry. *Toxicology and Industrial Health, 36*(9), 728–735. Retrieved from https://doi.org/10.1177/0748233720964631

CBI. (2018). What are the opportunities for luxury tourism from Europe? *CBI*. Retrieved from https://www.cbi.eu/market-information/tourism/luxury-tourism

Chabrak, N., Craig, R., & Daidj, N. (2016). Nouveau riche, old guard, established elite: Agency and the leadership of Vivendi Universal. *Leadership, 12*(4), 398–419. Retrieved from https://doi.org/10.1177/1742715014565442

Chaleila, W. (2019). The collapse of the American upper-class collective identity: Capitalism and the nouveaux riches in Edith Wharton's The House of Mirth. *Cogent Arts & Humanities, 6*(1), 1594515. Retrieved from https://doi.org/10.1080/23311983.2019.1594515

CLIA (2021). *2021 State of the cruise industry outlook*. Retrieved from https://cruising.org/-/media/research-updates/research/2021-state-of-the-cruise-industry_optimized.ashx

Collins Dictionary. (2021). *Masstige definition and meaning | Collins English Dictionary*. Retrieved from https://www.collinsdictionary.com/dictionary/english/masstige

Copeland, C. (2011). *Cruise ship pollution: Background, laws and regulations, and key issues*. Congressional Research Service. Retrieved from https://crsreports.congress.gov/product/pdf/RL/RL32450/14

Cruise Market Watch. (2021a). *Capacity | Cruise Market Watch*. https://cruisemarketwatch.com/capacity/

Cruise Market Watch. (2021b). Financial breakdown of typical cruiser. *Cruise Market Watch*. Retrieved from https://cruisemarketwatch.com/financial-breakdown-of-typical-cruiser/

Cruisewatch. (2021). *Highest cruise ships passenger-crew ratio*. Cruisewatch. Retrieved from https://www.cruisewatch.com/cruise-guide/rankings/ships-passenger-crew-ratio

Eskins, J. (2021). How cruise lines are using new itineraries and small cruise ships to attract eco-minded travelers. *Condé Nast Traveler*. Retrieved from https://www.cntraveler.com/story/small-cruise-ships-are-more-sustainble-and-on-the-rise

FCCA. (2006). *Cruise industry overview—2006*. Retrieved from https://www.f-cca.com/downloads/overview2006.pdf

Fearnley, H. (2021). What is the difference between Emporio Armani and Armani exchange? *Tessuti Blog*. Retrieved from https://blog.tessuti.co.uk/what-is-the-difference-between-emporio-armani-and-armani-exchange/

Global Industry Analysts. (2021). *Global yacht industry industry*. Retrieved from https://www.reportlinker.com/p02284917/Yacht-Industry.html?utm_source=GNW

Grand View Research. (2021a). *Yacht market size, share & trends analysis report, 2021–2028*. Retrieved from https://www.grandviewresearch.com/industry-analysis/yacht-market

Grand View Research. (2021b). *Yacht market size worth $12.16 billion By 2028*. Retrieved from https://www.grandviewresearch.com/press-release/global-yacht-market

Hancock, A. (2020). Cruise industry's billion-dollar question: What to do with the ships? *Financial Times*. Retrieved from https://www.ft.com/content/b0d09282-1559-4169-a3ec-92c12ebfb78e

Holland America Line. (2021). *Entertainment*. Retrieved from https://www.hollandamerica.com/en_US/onboard-activities/cruise-entertainment.html

Huang, J., & Hsu, C. H. C. (2009). Interaction among fellow cruise passengers: Diverse experiences and impacts. *Journal of Travel & Tourism Marketing, 26*(5–6), 547–567. Retrieved from https://doi.org/10.1080/10548400903163103

Ingram, R. (2021). The 2021 superyacht trends you need to know. *Forbes.* Retrieved from https://www.forbes.com/sites/rachelingram/2021/01/27/the-2021-superyacht-trends-you-need-to-know/

Ioannides, D., & Gyimóthy, S. (2020). The COVID-19 crisis as an opportunity for escaping the unsustainable global tourism path. *Tourism Geographies, 22*(3), 624–632. Retrieved from https://doi.org/10.1080/14616688.2020.1763445

Ito, H., Hanaoka, S., & Kawasaki, T. (2020). The cruise industry and the COVID-19 outbreak. *Transportation Research Interdisciplinary Perspectives, 5*, 100136. Retrieved from https://doi.org/10.1016/j.trip.2020.100136

Kapferer, J.-N., & Bastien, V. (2012). *The luxury strategy: Break the rules of marketing to build luxury brands.* Kogan Page Publishers.

Kapferer, J.-N., & Bastien, V. (2017). *The specificity of luxury management: Turning marketing upside down* (pp. 65–84). Retrieved from https://doi.org/10.1007/978-3-319-51127-6_5

Kapferer, J.-N., Klippert, C., & Leproux, L. (2014). Does luxury have a minimum price? An exploratory study into consumers' psychology of luxury prices. *Journal of Revenue & Pricing Management, 13.* Retrieved from https://doi.org/10.1057/rpm.2013.34

Kraus, M. W., Park, J. W., & Tan, J. J. X. (2017). Signs of social class: The experience of economic inequality in everyday life. *Perspectives on Psychological Science, 12*(3), 422–435. Retrieved from https://doi.org/10.1177/1745691616673192

Levius, T. (2020). Why your next cruise should be a yacht cruise. *Condé Nast Traveler.* Retrieved from https://www.cntraveler.com/story/why-your-next-cruise-should-be-a-yacht-cruise

Lew, A. A. (2018). Why travel?—Travel, tourism, and global consciousness. *Tourism Geographies, 20*(4), 742–749. Retrieved from https://doi.org/10.1080/14616688.2018.1490343

Lipcon, Margulies, Winkleman (2021). *Number of cruise lines.* LM&W, PA. https://www.lipcon.com/frequently-asked-questions/many-cruise-lines-world/

LVMH. (2021). Press, media, exceptional places and arts—Other activities. *LVMH.* Retrieved from https://www.lvmh.com/houses/other-activities/

Mallet, V. (2010). Environment: Demand for more sustainable boats and equipment grows. *Financial Times.* Retrieved from https://www.ft.com/content/d3fd1762-c482-11df-bc11-00144feab49a

Manley, B., Elliot, S., & Jacobs, S. (2017). Expedition cruising in the Canadian arctic: Visitor motives and the influence of education programming on knowledge, attitudes, and behaviours. *Resources, 6*(3), 23. Retrieved from https://doi.org/10.3390/resources6030023

Mitruka, K., & Wheeler, R. E. (2008). Cruise ship travel. *Travel Medicine*, 351–360. Retrieved from https://doi.org/10.1016/B978-0-323-03453-1.10034-3

Montigneaux, R. (2019). *2019 global order book*. Retrieved from https://www.boatinternational.com/yacht-market-intelligence/luxury-yachts-on-order/2019-global-order-book--39251

Moreau, C. P., Prandelli, E., Schreier, M., & Hieke, S. (2020). Customization in luxury brands: Can Valentino get personal? *Journal of Marketing Research, 57*(5), 937–947. Retrieved from https://doi.org/10.1177/0022243720943191

Mortelmans, D. (2005). Sign values in processes of distinction: The concept of luxury. *Semiotica, 2005*(157–1–4), 497–520.

P&O. (2021). Entertainment on-board. *P&O cruises*. Retrieved from https://www.pocruises.com/onboard-activities/entertainment

Princess Cruises. (2021). *Princess cruises: Onboard entertainment*. www.Princess.Com. Retrieved from https://www.princess.com/ships-and-experience/onboard-experience/entertainment/

Radon, A. (2012). Luxury brand exclusivity strategies—An illustration of a cultural collaboration. *Journal of Business Administration Research, 1*. Retrieved from https://doi.org/10.5430/jbar.v1n1p106

RCCL. (2021). Symphony of the seas fact sheet. *Royal Caribbean Press Center*. Retrieved from https://www.royalcaribbeanpresscenter.com/fact-sheet/31/symphony-of-the-seas/

Regent Cruises. (2020). *Regent seven seas cruise ships*. Retrieved from https://www.regentcruises.com/cruise-lines/regent-seven-seas-cruises-ships.html

Regent Seven Seas. (2021). *Regent seven seas homepage*. Retrieved from https://www.rssc.com/ships/seven_seas_splendor

ReportLinker. (2020). Global yacht industry industry. *GlobeNewswire News Room*. Retrieved from https://www.globenewswire.com/news-release/2020/03/11/1998982/0/en/Global-Yacht-Industry-Industry.html

Rizzo, C. (2021). Carnival says it already has more bookings for 2022 than it had in 2019. *Travel + Leisure*. Retrieved from https://www.travelandleisure.com/cruises/carnival-cruises-booking-numbers-2022

ROL Cruise. (2021). *Crystal Espirit cruises 2021: Crystal cruise ships. ROL Cruise*. Retrieved from https://www.rolcruise.co.uk/cruise-lines/crystal-cruises/crystal-esprit

Royal Caribbean. (2021). Cruise shows & live entertainment. *Royal Caribbean Cruises*. Shows. Retrieved from https://www.royalcaribbean.com/gbr/en/experience/cruise-shows-and-entertainment

Seabourn. (2021a). *Entertainment*. Retrieved from https://www.seabourn.com/en_US/experience/onboard-activities/entertainment.html

Seabourn. (2021b). *Seabourn homepage*. Retrieved from https://www.seabourn.com/en_US.html

Sevinc, F., & Güzel, T. (2018). Sustainable yacht tourism practices. *Management & Marketing, XV*, 61–76.

Silverseas. (2021). *Silverseas homepage*. Retrieved from https://www.silversea.com/

Soares, P., Rito, T., Trejaut, J., Mormina, M., Hill, C., Tinkler-Hundal, E., Braid, M., Clarke, D. J., Loo, J.-H., Thomson, N., Denham, T., Donohue, M., Macaulay, V., Lin, M., Oppenheimer, S., & Richards, M. B. (2011). Ancient voyaging and Polynesian origins. *The American Journal of Human Genetics, 88*(2), 239–247. Retrieved from https://doi.org/10.1016/j.ajhg.2011.01.009

Solent University. (2021). *The superyacht industry. Maritime industry. Warsash Maritime*. Retrieved from https://maritime.solent.ac.uk/maritime-industry/superyachts

Spence, E. (2014). Unraveling the politics of super-rich mobility: A study of crew and guest on board luxury yachts. *Mobilities, 9*(3), 401–413. Retrieved from https://doi.org/10.1080/17450101.2014.946772

Steele, R. (2021). *Coronavirus—Yacht charter & superyacht news*. Retrieved from https://www.charterworld.com/news/category/coronavirus

Syriopoulos, T., Tsatsaronis, M., & Gorila, M. (2020). The global cruise industry: Financial performance evaluation. *Research in Transportation Business & Management*. Retrieved from https://doi.org/10.1016/j.rtbm.2020.100558

Travel Weekly. (2021). *Forging fulfilment: Cruise lines bolster enrichment programs*. Retrieved from https://www.travelweekly.com/Cruise-Travel/Forging-fulfillment-Cruise-lines-bolster-enrichment-programs

Uğur, N. G., & Akbıyık, A. (2020). Impacts of COVID-19 on global tourism industry: A cross-regional comparison. *Tourism Management Perspectives, 36*, 100744. Retrieved from https://doi.org/10.1016/j.tmp.2020.100744

Warwick University. (2021). *Research from a practitioner perspective*. Retrieved from https://warwick.ac.uk/fac/soc/ier/ngrf/effectiveguidance/research/practitioner_perspective/

Williamson, J. (2021). *Why 2021 is great for small ship cruising*. Retrieved from https://www.cruisecritic.co.uk/articles.cfm?ID=5941

Windstar. (2020). How many people on a cruise ship | Windstar. *Windstar Cruises Travel Blog*. Retrieved from https://www.windstarcruises.com/blog/how-many-people-on-cruise/

Wondirad, A. (2019). Retracing the past, comprehending the present and contemplating the future of cruise tourism through a meta-analysis of journal publications. *Marine Policy, 108*, 103618. Retrieved from https://doi.org/10.1016/j.marpol.2019.103618

Zolfani, S. H., Sedaghat, M., Maknoon, R., & Zavadskas, E. K. (2015). Sustainable tourism: A comprehensive literature review on frameworks and applications. *Economic Research-Ekonomska Istraživanja, 28*(1), 1–30. Retrieved from https://doi.org/10.1080/1331677X.2014.995895

Part II

Perspectives on Practice

5

On Board a Luxury Vessel: Demystifying the Daily Operations

Victoria Gladkikh

Introduction

Yachting tourism is a type of special interest tourism (SIT) which has not been clearly defined (Mikulić et al., 2015). However, its various forms are interchangeably used to describe the nature of the activities associated with yachting tourism. Terms such as nautical tourism, leisure boating, marina tourism are often used as synonymous (Mikulić et al., 2015). The luxury yachting industry offers another form of yachting tourism distinguished by privately owned superyachts more than 30 m in length which are professionally crewed. An average 47 m motor yacht costs c.30 million Euros to buy new from a reputable European shipyard, and the operational and running costs can be estimated as millions (Adamczyk, 2015). Thus, this type of leisure activity is exclusively accessible to the

V. Gladkikh (✉)
Independent Luxury Yacht Practitioner, Dunkeld, UK

T. Gladkikh et al. (eds.), *Luxury Yachting*,
https://doi.org/10.1007/978-3-030-86406-4_5

super-rich (Spence, 2016) and the details of luxury yacht operations are not widely known to the general public.

There is some information currently available online in media articles (www.thetimes.co.uk) (Agnew, 2020) and on reality TV shows, such as Below Deck (shown on Netflix, Amazon Prime, Apple TV and Bravo TV) but this information is not conclusive. This chapter will discuss the claims made about the industry in the media and then go on to provide practical examples of real-life situations on yachts. It will systematically explain each department on board and elaborate on how the crew facilitate guests with a trip of a lifetime within this pinnacle of luxury tourism. Ultimately, this chapter will conclude that the industry's unique orientation attracts many superficial onlookers who are quick to assume much about the 'upstairs, downstairs' culture of yachting and luxury tourism in general.

In reality, the luxury yachting industry is not nearly as exaggerated as the media portrays it to be. It brings local, shore-based tourism industries a myriad of benefits, specifically to popular ports of call and coastal areas worldwide. Benefits include employment creation and mass capital inflow, as well as income generation for shore-based industries that provide support for the luxury vessels that facilitate this sector of luxury tourism. Yachting is an extremely professional and lucrative sector of luxury tourism which is rarely considered in many studies. A simple search on Amazon Books shows that there are only three books available to purchase about the luxury yachting industry. These books are relatively recent, they were written between 2015 and 2018, which shows that there is interest in this subject but clearly there are not enough physical publications available to provide readers with a wide range of sources.

This chapter aims to fill a gap in luxury tourism literature while emphasizing the need for a deeper understanding of this sector that has currently not received the desired level of academic attention. It is written solely from a practitioner's perspective using experiential evidence approach which builds on the professional experience and judgment of practitioners (Barends et al., 2014), with primary evidence generated through own first-hand experiences. Additionally, it utilizes subjective personal introspection (SPI) (Holbrook, 1995) into the findings of telephone interviews conducted with two crew who are currently working in this sector.

Media and TV

Despite the hardships that have been plaguing the world within the last two decades, such as the financial crisis of 2008 and more recently, the outbreak of COVID-19, the yachting industry has managed to remain confidently resilient. As an exclusive industry, its self-sufficient nature benefits from large sources of private funding which makes it continuously stable. It is not an industry that suffers hardships or losses, the world of luxury yachting is a world of its own. General public knowledge of the industry often comes from opinion articles written by journalists claiming to know what it is really like to work on board (www.thetimes.co.uk) (Agnew, 2020) and reality TV shows, such as Below Deck, which boast the ability to capture 'reality' vividly and let viewers dip a toe into yacht life. Below Deck has now been flanked by two other related shows: Below Deck Mediterranean and Below Deck Sailing Yacht, all of which have helped to greatly increase viewer interest in yachting but have not necessarily provided the most accurate depiction.

The escapism that Below Deck provides to viewers has helped to exponentially grow the show's popularity, especially recently when many viewers have been trapped in a lockdown during the COVID-19 pandemic. The most recent season of Below Deck Mediterranean aired from 1 June 2020 until 26 October 2020, these exact dates blanketing a period when many individuals found themselves in periods of quarantine or lockdown worldwide. The correlation between this period and the huge increase in the show's viewer numbers is obvious. The 2020 season achieved record after record and became the highest rated series in the entire history of the franchise, reaching nearly 1.9 million people during overnight viewing at one point (www.cheatsheet.com) (Ragusa, 2020).

Below Deck portrays life on board for the crew to be relaxed and lavish, with crew often seen engaging in partying with guests or walking around shirtless on a guest trip. According to a yachting crew member, Deckhand Steve Wright 'A deckhand will never take his shirt off; that's completely false' (www.the-triton.com) (Serio, 2016). The lack of respect for hierarchy is also striking in the Below Deck series. Captains are often too lenient with the crew and are not disciplining unprofessional behaviour such as loud, bolshy arguments, especially with guests on board. First Officer Cornelis De Necker of M/Y Mia Elise agrees: 'It's terrible (and) seems scripted. The chit-chat to senior officers is not how it works'. At the end of the day, yacht crew are expected to embody the seven-star service that this sector of luxury tourism boasts. Perhaps an area of yachting that Below Deck does show well is the buzzing energy that crew radiate, stemming from the extremes of coffee and adrenaline-fuelled tiredness during a trip to utter elation once the guests have departed, with tip money in hand and a day off within reach: 'the zigzag of frantic, round-the-clock shifts followed by sudden reprieves between charters… fosters a work hard, party hard atmosphere' (www.nytimes.com) (Weaver, 2020).

Articles such as Megan Agnew's piece for The Sunday Times, draw attention to the extreme negatives of the industry. The sombre tone conveys an image of the very worst in the general public that are reading the article, who often know very little about the industry itself (other than maybe what they see on Below Deck, glamorized, scripted and what actual yacht crew call 'career suicide'). In addition, this article's polarizing ideas continue to distort the practical and professional essence of yachting by making exaggerated claims of 'Swarovski-encrusted anchors' for example, which cannot be backed up by physical evidence. Simply Google search 'Swarovski encrusted anchor, yachts' and all the results that are shown are merely jewellery. From a standpoint of personal experience in this industry, these exaggerated claims beg the question of the credibility of the article.

As a result, there is a clear need for a deeper and more realistic understanding of the industry and how it operates. It is undeniable that choosing to work on a yacht is a taxing career to say the least, but this is the crux of the industry—hard work comes with great reward. Perks come in all shapes and sizes, from essentially 'free travel' to a large paycheck and living in luxury accommodation with a crew that one may come to love like extended family. Sometimes crew may even benefit from being 'guinea pigs' for the Chefs to try out their ideas for guest meals. Lesser known is the fact that juggling work and life on a yacht is not suited to everyone as living with the same crew while working, eating and also relaxing together in the tight confines of a yacht can be a real challenge.

When speaking about yachting or about the role of a crew member, it is often hard to generalize because no two boats are ever the same. Each yacht has its own itinerary, goals and culture, but there are many different paths towards the delivery of the luxury tourism service. The discussion below focuses on the three departments on a 60–80 m yacht, namely Deck, Interior and Engineering, and explores the main themes of their daily operations.

The Deck Department

The Deck Department is responsible for maintaining the exterior of the vessel (otherwise known as the 'decks'), moving the vessel from A to B at the owners' request, maintaining and operating the tenders (a smaller boat with a motor, usually stored inside the yacht's own garage, used to transport guests, crew, provisions and to facilitate water sports activities) and any water sports equipment, as well as implementing and enforcing safety protocol on board. Deck hierarchy on a 60–80 metre vessel is organized in a traditional structure, the chain of command starting with the Captain, followed by the Senior Management, a Chief Officer and Second Officer, then the general employees, the Bosun and Deckhands (crewing numbers and hierarchical structure of course directly correlate with the size of the vessel).

Guests generally spend more time off the boat than on it. 'Owners or guests are typically not on the boat all that much, so most of the time the crew has the boat to themselves' (www.businessinsider.com) (Hoffower, 2020). A common misconception is that crew are not required to work when the guests are not on board. In reality, a work list is infinite on a yacht and the crew are required to work year-round, much like a normal job. With no guests, working hours are reduced to a regular 8–5 and work is less demanding. Day-to-day maintenance of the exterior of a yacht usually consists of washing down the whole exterior once a week, solely to preserve the quality of the paint by preventing dirt and sand from accumulating on it and abrading it. This work is a Deckhand's 'bread and butter'. Other projects that are messier and require longer periods of time to complete are usually reserved for longer 'off' periods when the yacht is taken into a shipyard for refit. This usually happens once a year but can, of course, be more frequent depending on itinerary and budget constraints. These longer tasks include but are not limited to exterior painting and paint repairs of chips, scratches or imperfections, varnishing capping rails and furniture, carpentry jobs, replacing or laying new caulking as well as rust and metal works. Additionally, there are numerous smaller routine jobs such as whipping and splicing of lines, organization and cleaning of deck spaces as well as inventorying the deck stores of paint, tools, chemicals, varnishes, polishes, acids and hardware, to name a few. Stocktaking and reordering is usually the job of the Bosun.

When preparing for guest arrival, deck work starts to become more demanding with time pressures being introduced. The Bosun and Deckhands focus predominantly on washing and scrubbing the teak decks (usually using a machine to remove the top 'grey' layer of sun damaged wood on the decks, then brightening the newly exposed wood). Then detailing, which consists of polishing and protecting all the stainless steel on board (handrails, staircases, lights, anchors), removing any marks on the exterior paint that did not come off during the wash (usually using alcohol), followed by protecting the paint, polishing window glass and putting in place the exterior furniture and plants as per the 'guest set up'. Final preparation details involve checking and detailing the beach gear and water sports equipment, particularly ensuring that all the 'toys' are working correctly.

The Captain liaises with the guests to plan their arrival and departure points and trip itinerary. His responsibilities are booking relevant marinas, organizing fuelling arrangements, dealing with budgets, expenditure and crewing requirements. Depending on the itinerary chosen, it is the responsibility of the navigational Officer (usually the Second

Officer) to create appropriate passage plans and monitor weather forecasts. The Chief Officer focuses on safety, specifically the Safety Management System (SMS) to ensure that the yacht is conforming to legalities and has appropriate safety procedures in place before the guests step on board. Routine daily, weekly and monthly safety checks are completed by the Deck crew. The Captain and Officers also enforce routine drills and crew safety training.

Once the guests have arrived on board, workload increases considerably. A standard working day 'on trip' is anything upwards of 14 hours and days off are not permitted until the guests depart. The Deck crew are responsible for keeping the vessel exterior looking pristine. An anonymous First Mate who was interviewed for a Business Insider article is correct to point out that 'they (guests) expect it to look like no one has touched it, so any rain or dew, water spots or salt spray has to be continuously cleaned, not to mention having to constantly wipe stainless (steel) and glass when guests move around the boat. They leave smudges and fingerprints everywhere' (www.businessinsider.com) (Hoffower, 2020). This includes vacuuming the decks, fluffing pillows and collecting rogue, wet towels then sending them to the laundry. The Deck crew are responsible for all tender runs, either transporting guests from the yacht to town/the beach/lunch/other yachts or collecting provisions from the shore and delivering them to the vessel.

The Deck crew are tasked with navigational operations each day of the guest trip, either moving anchorages or coming in and out of port as per guest request, all the while practising good seamanship, watching the weather, tides, currents, advising guests on safe jet ski areas and using the tender to build excitement and adrenaline on this luxury holiday with water sports such as wake boarding, water skiing and tubing. Generally, the Deck crew do not communicate with the guests too often unless they are doing water sports or tender runs, in which case communication is usually verbal and casual with any guest requests to be fulfilled immediately and with no hesitation. Usually, the Captain will liaise with the guests throughout the day to verify that they are enjoying their holiday and to find out about any trip itinerary changes—it is normal for the itinerary to change several times each day which means that the crew must work quickly and be prepared for anything that the guests request—nothing is impossible in the world of luxury yachting.

The Interior Department

The Interior Department is responsible for the comfort and entertainment of guests, service of meals, snacks and drinks, cleanliness of all interior areas and the laundry. This department is sub-divided into smaller areas of expertise, comprising of service, housekeeping, laundry and the galley. The Chief Stewardess leads the Interior, with assistance from the Head of Service and Head of Housekeeping. The hierarchical structure continues with Stewardesses who rotate between service, housekeeping and sometimes laundry duties, when there is no designated Laundry person on board. The Purser is also a member of the Interior team, usually working in the Bridge and assisting the Captain with guest excursions, accounts, vessel and crew movements. The galley team is a separate branch of the Interior Department made up of a Head Chef and Sous Chef.

As standard across all yachts, Interior daily duties include cleaning the bridge, all crew areas and doing crew laundry. The Galley department will prepare hot lunch and dinner for the crew and sometimes in addition will make breakfast or something sweet for afternoon tea break; crew are otherwise left to fend for themselves in a crew mess stocked with extensive snacks, hot and cold drinks. In this sense, yacht crew members are extremely fortunate to have all their food prepared by professional Chefs daily. This eliminates the need of having to prepare your own meals to take to work, as many people who work ashore do. It also ensures that crew members are fuelling their bodies with proper nutrients from a hot meal, which in turn assists them with energy for their physically demanding line of work. The Chefs will usually go provisioning for crew food once a week. Without guests on board, once the daily duties are complete, Stewardesses will 'dust vac' (the action of vacuuming up dust from all surfaces using the round, fluffy head attachment on a vacuum cleaner) and wipe all surfaces in the guest interior to eliminate dust build-up from the air conditioning. Other important tasks include inventorying every cupboard and area in the yacht's interior, as well as following the monthly cleaning maintenance schedule which details a specific area of the Interior that gets 'extra polished' each month. By living on the fringe of luxury in spotlessly clean crew accommodation, crew are provided with a good quality of life, even when living in relatively small crew quarters.

Akin to military precision, a high level of organization is expected from yacht Stewardesses. After speaking with a Head Housekeeper on an 82 m yacht, the extent of preparations and detail prior to guest arrival became evident: 'we have pre-charter checklists, for 20 days before a trip, 10 days, 5 days and 3 days before (After a crossing to a new destination), everywhere needs to be vacuumed again, everywhere needs to be cleaned. The boat needs to be set up to perfection. We measure the distance between an object and the edge of the table, everything is one cm. Most things are measured in place so that they are perfect. We have set up photos of exactly where things were the last time the owners were on board so that everything is exactly the same when they walk on'. Every detail is meticulously planned in advance and recorded, from sheet change schedules and theme night schedules to menus and service schedules. The most stressful time in the role of a Stewardess is in the lead up to the start of a guest trip. From the Head Housekeeper's experience, once the guests are on board, everything flows because everything has already been prepared. The galley department will liaise with the Owners, the Captain and the Chief Stewardess to create menus for the trip, usually including some local cuisine for fun theme nights. Provisions will be ordered accordingly, with crew and guest food distinctly separated. The Chief Stewardess will also order wines and spirits for cocktails, as per the guests' preferences.

Once the guests are on board, the momentum of hard work keeps going: 'long days, little rest, expectations to perform at the highest level of service and not cracking under pressure' (www.businessinsider.com) (Hoffower, 2020). To the right individual, being rushed and put under pressure to do a really good job is one of the most exciting aspects of the role. The success of the trip is very much dependent on the guests and their interests, to which the role of a Purser is key: to be a liaison with the guests and facilitate any requests regarding tourism excursions. Preferences are hard to generalize as all guests/owners have different ideas for how they want to spend their luxury holiday. Excursion tourism activities, such as seeing historic landmarks, are enjoyed mostly by charter guests that have hired the yacht to travel. However, to most guests, especially private vessel owners, the extent of tourism activities only stretches to shopping and restaurant meals.

From first-hand experience, most guests prefer to use the facilities on board such as the gym or spa and spend their time doing water sports before enjoying sunset cocktails and a meal prepared by the Chefs to suit their tastes. The Head Housekeeper agreed with this statement and added that: 'the guests would rather stay on the boat and go off for 20 mins to go to Fendi, not to go and look at a volcano'. The clientele that the luxury tourism industry naturally attracts means that guests do not have shared values with other types of tourists, such as sightseeing on land. Instead, they specifically choose a yacht to be able to easily do water sports and spend their holiday in secluded locations on the water. The word 'luxury' appeals to high-paying tourists who want to take advantage of onboard spa treatments, personalized service and the freedom to do anything they want to on their own private yacht. To these people, exploring crowded tourist sites does not reflect their idea of a dream holiday.

The Engineering Department

The responsibilities of a yacht's Engineering department is to ensure that all of the mechanical, electrical, electronic, hydraulic and structural systems on board are operating correctly and are maintained and repaired as required so that the yacht can operate freely and is able to use all of her machinery areas. On a vessel between 60–80 m, there would commonly be a Chief Engineer and Second Engineer, in some cases a Third Engineer and/or an Electronic Technical Officer (ETO). Successful engineers have a good breadth of technical knowledge as well as a sharp attention to detail. The ability to spot minor changes or differences in the engine room, such as a small leak in a bilge, could potentially point to a larger, hidden issue and could be the difference between a quick repair or a full-blown disaster.

A lot of work in the engine room is based around planned maintenance and scheduled checks, with a lot of yachts relying on yacht management software, such as Idea, an 'industry leading yacht management software…(which) supports…execution of key operational tasks on board a vessel' (www.idea-yacht.com) (Idea Yacht, 2020) to ensure that there is no human error with maintenance scheduling. According to a Second Engineer on a 48 m yacht, every task on board is organized into a maintenance schedule and will have a date or time for completion. Once you have done the task, you sign it off and the schedule will reset until it needs to come up again. This highlights the systematic and organized nature of a yacht Engineer's line of work. Much like the Stewardess', everything is pre-planned and organized, however with Engineering the stakes are much higher. In this department, planning is done to eliminate the risk of human error when working with critical machinery for the safety of life and the vessel.

The Second Engineer shared that when preparing for a guest trip, Engineers go through all the safety systems, fire alarms and fire procedures for the vessel so that they know by law that everything is working correctly.

In addition to this, the Engineering department will perform any maintenance or repairs on tenders or mechanical water sports toys to ensure that everything runs smoothly while the guests are on board. Tenders and water sports equipment are heavily used so if they were to fail during a trip, it would result in the quality of the luxury tourism experience deteriorating. Prior to a guest trip, the Interior crew usually do an Interior walk-through to make sure that everything is working in the guest areas. From the Second Engineers experience, the main things that could fail in the guest interior are lightbulbs. Fuelling also needs to happen prior to guest arrival. Together with the Bridge team, the Engineers would consult the itinerary and passage plans for the duration of the upcoming guest trip and then perform calculations to ascertain how much fuel and oil would need to be taken on for the duration of the trip, with extra leeway in case of an (inevitable) change in plans or in an emergency.

During guest trips, Engineers must be on call 24/7 in case of any issues or when the yacht is underway to a new location. When at sea, many systems on board are safety critical so breakdowns and failures offshore bring a whole new set of challenges and risks. This puts pressure on the Engineering department to keep all systems working to eliminate down time and negative impacts on the itinerary and guest expectations (www.wilsonhalligan.com, 2020) (Wilson, 2020). The Second Engineer highlighted other priorities of Engineering when the guests are on board, including ensuring that the Internet is working as fast as possible, the air conditioning is getting down to the requested temperature, the crew radios and communications are working perfectly as well as the guest entertainment centres. If that is all working, the guests are happy.

Compared to the service-orientated work of the Deck and Interior departments, the Engineers are more technically focused due to the nature of their profession. From the Second Engineer's experience, if the Engineering department can do a whole guest trip without seeing the guests, then they know that they have done their job well because nothing was broken. When the guests do see the Engineers, they panic a little and think that something is wrong, so they generally try and stay out of the way. In addition, one of the biggest roles of a good Engineer is to relieve some of the stress from other crew members whose jobs, at the time of a guest trip, are the most under pressure. The Second Engineer elaborated on this point and shared that one of the challenges of his role is keeping the morale of the crew going: 'we try to keep everyone happy, keep laughing and also try to keep ourselves sane! During the season, it is our (the engineering department's) quietest time, hoping that nothing breaks and trying to keep busy'.

This often results in the Engineers helping other departments, for example assisting the Deck crew to launch tenders and do waste disposal ashore or helping the Interior crew by taking over the crew mess and packing it up after crew mealtimes. Successful yacht crew members understand that everyone, regardless of which department they are in, is working towards a common goal—the delivery of a luxury tourism experience. In contrast to Below Deck, where the crew often make life difficult for themselves by arguing and pulling each other apart, real yachting is about a collective approach to working, rather than the individual differences. Important qualities of crew members are to be respectful of different cultures and of each other, as well as being able to work as a team under the umbrella of shared values. This really shows the essence of the togetherness of yachting and how, contrary to the picture painted by the media, the spirit of collaboration really goes a long way in an industry like this.

To conclude, this chapter has presented the image of the luxury yachting industry that is being portrayed in the media and has offered a real-life account of the main responsibilities of the Deck, Interior and Engineering departments. By offering these opposing aspects on the same industry, this chapter has emphasized that there is a need for a deeper understanding of this sector and how it operates so that information across the board is cohesive and fills the gap in luxury tourism literature.

This chapter has also demonstrated that each department on board has varying levels of direct exposure to the experience of luxury tourism. The Interior and Galley have the highest level of exposure, organizing excursions and meals featuring local cuisines; followed by the Deck department who organize travel itineraries and water sports activities. On the contrary, the Engineering department are practically and mechanically orientated, trying to limit guest interaction as much as possible. However, without these departments collaboratively working together, a smooth and professional luxury yachting experience cannot be guaranteed.

Without the technical knowledge of the Engineering department, the yacht would not run and without the Captain's and Deck Officer's navigation skills, there would be no opportunity to cruise to secluded bays and show the guests a truly unique experience, otherwise inaccessible to the general public. It is this togetherness that really highlights the contemporary realities and professionalism of this sector, something that can be misrepresented by media and reality TV shows that popularize a one-sided portrayal of the industry. Ultimately, this points to a need for more publications like this one, with the sole purpose of deepening a true understanding of how this niche sector operates and how it really embraces the demands of tourists, professionals and the luxury tourism industry itself.

Certification and Training

The development of yachting tourism, as well as luxury yachting tourism, has significant economic benefits such as increased tourism investment, lower seasonality of tourist demand and more attractive destination image. Another notable impact is employment possibilities (Mikulić et al., 2015). This section aims to provide basic guidance for anyone interested in embarking on a career in the yachting industry. The requirements below are aimed towards people who are in the initial stages of a yachting career and who are looking to gain the basic qualifications that are necessary for simply working on board a vessel, before beginning to consider specializing into one of the aforementioned departments.

Below are two compulsory requirements for every professional job in the industry:

International Convention on Standards of Training, Certification and Watchkeeping (STCW)—STCW certification was created to promote safety of life, safety of the vessel and to protect the marine environment. It establishes internationally accepted standards of training and certification of seafarers, ensuring that crew are qualified and fit for duties at sea. The course is five days long and covers basic training in personal survival techniques, firefighting and fire prevention, personal safety and social responsibility, elementary first aid and proficiency in security awareness.

UK flag-state medical examination (ENG1)—The ENG1 is a medical examination set by the Maritime and Coastguard Agency (MCA) to assess whether individuals are fit to work on a seagoing vessel and perform duties at sea. This medical examination is only performed by MCA-approved Doctors. A full explanation of the medical exam can be found online when searching 'MCA ENG1'.

As always, there is an abundance of information that can be accessed online regarding maritime training and certification for further knowledge such as:

MCA Training and certification: detailed information—https://www.gov.uk/topic/working-sea/training-certification

Seafarer skills and training—https://www.gov.uk/seafarer-skills-and-training

Bluewater Yachting—https://www.bluewateryachting.com/crew-placement/yacht-crew

Superyacht UK—https://www.superyachtuk.com/careers/

List of MCA Approved ENG1 Doctors (UK based)—https://www.gov.uk/government/publications/mca-approved-doctors-uk-based

List of MCA Approved ENG1 Doctors (overseas)—https://www.gov.uk/government/publications/mca-approved-doctors-overseas

Acknowledgements Many thanks to Alice and Harry for giving their time and for providing their insightful practice-based inputs to this chapter (telephone interviews, 10 November and 18 November 2020 respectively).

References

Adamczyk, A. (2015, April 8). *How much does a superyacht really cost?* Forbes. Retrieved March 21, 2021, from https://www.forbes.com/sites/aliciaadamczyk/2015/04/08/how-much-does-a-superyacht-really-cost/?sh=4d6c104936bc

Agnew, M. (2020, January 26). The secretive world of superyachts: What is it really like to work onboard? *The Times*. Retrieved November 18, 2020, from https://www.thetimes.co.uk/article/the-secretive-world-of-superyachts-what-is-it-really-like-to-work-onboard-b5rgzv8mh

Barends, E., Rousseau, D. M., & Briner, R. B. (2014). *Evidence-based management: The basic principles*. Centre for Evidence-Based Management.

Hoffower, H. (2020, March 9). 9 superyacht crew members share what it is really like working for a billionaire on board. *Business Insider*. Retrieved November 11, 2020, from https://www.businessinsider.com/yacht-crew-

reveal-what-its-like-to-work-on-superyacht?r=US&IR=T#they-can-be-more-down-to-earth-than-expected-but-their-spontaneity-can-be-frustrating-9
Holbrook, M. B. (1995). *Consumer research: Introspective essays on the study of consumption.* Sage Publications.
Idea Yacht. (2020). *Operational software.* Retrieved November 12, 2020, from https://idea-yacht.com/yacht-software-by-category/#OperationaSoftware
Mikulić, J., Krešić, D., & Kožić, I. (2015). Critical factors of the maritime yachting tourism experience: An impact-asymmetry analysis of principal components. *Journal of Travel and Tourism Marketing, 32*(sup1), S30–S41.
Ragusa, G. (2020, December 24). *'Below Deck' Bravo ratings domination actually started with 'Below Deck Med'.* Retrieved March 8, 2021, from https://www.cheatsheet.com/entertainment/below-deck-bravo-ratings-domination-actually-started-with-below-deck-med.html/
Serio, T. (2016, November 6). *Triton today survey: Does "Below Deck" portray yachting's reality?* The Triton. Retrieved November 11, 2020, from https://www.the-triton.com/2016/11/triton-today-survey-does-below-deck-portray-yachtings-reality/
Spence, E. (2016). Eye-spy wealth: Cultural capital and 'knowing luxury' in the identification of and engagement with the superrich. *Annals of Leisure Research, 19*(3), 314–328.
Weaver, C. (2020, June 29). The reality behind 'Below Deck': Filming Bravo's hit reality franchise about yachts and yachties is a complex logistical operation. *The New York Times.* Retrieved November 10, 2020, from https://www.nytimes.com/2020/06/29/style/below-deck-bravo.html
Wilson, H. (2020, January 30). *How do I become a yacht engineer—Qualifications and requirements.* Retrieved November 12, 2020, from https://www.wilsonhalligan.com/how-to-become-a-yacht-engineer/

6

Sustainability and Luxury Yachting: the Evolution of a New Sustainable Era

Lauren Ryburn

Introduction

Fuel consumption, air pollution, waste production... There are many contributing factors impacting the luxury yachting industry to be commonly viewed as being environmentally unsustainable or having a large 'environmental footprint'. With over 2312 40 metre + superyachts in the world and more being launched every year (www.superyachtecoindex.com, 2021), the size of this environmental footprint will only grow if there are no drastic sustainable improvements made. Fortunately, over the last few years, the industry has begun to change and many of the unsustainable practices previously carried out on board are now becoming obsolete or commonly frowned upon. This chapter is the author's personal evaluation and interpretation of the progression and

L. Ryburn (✉)
Independent Luxury Yacht Practitioner, Whangaparaoa, New Zealand
e-mail: lauren@thegreenstew.com

T. Gladkikh et al. (eds.), *Luxury Yachting*,
https://doi.org/10.1007/978-3-030-86406-4_6

evolution of sustainability within the luxury yachting industry from a practitioner's perspective.

Having worked within the industry as Stewardess and Sustainable Yachting Advocate for over seven years, I have become alarmingly aware of the unsustainable practices carried out on board a yacht. This is what influenced me to begin making eco-conscious changes to the practices I carried out every day as a Stewardess and to inspire the fellow crew members to do the same. Positive responses to these changes lead me to create an eco-conscious yachting blog and Instagram page called 'The Green Stew' to share with other crew some of these practices and insights. The platform promotes sustainability within the industry, shares eco tips and offers guidance and support to fellow crew wishing to make a difference within the industry. Over this time my blog has gained 1800 followers, the platform hosted and been involved in several environmental awareness collaborations and crew beach clean-ups around the world and have had the honour of being awarded the 'Acrew Environmental crew member of the year award' 2020. This has been a very rewarding experience and I believe the industry has made notable advancements in its eco endeavours since The Green Stew started in 2018.

Luxury yachting is a very unique and a relatively new industry compared to more established ones, and it is growing and evolving every day. One of the vital ways in which it has progressed is through development of the collective mindset around sustainability and environmental awareness. This chapter will discuss some of the reasons why I believe the progression of sustainability within the yachting industry was held back for some time and evaluate the factors which have contributed to the positive changes we see today throughout the industry.

The Slow Start to Sustainability within the Superyacht Industry

I believe that one of the key factors that delayed and held back the progression of sustainability within the industry at first was disconnect on different levels:

- Disconnect between yachts within the industry.
- Disconnect between the industry and the rest of the world.
- Disconnect between the luxury yachting industry and the collective mindset and understanding of its innate connection to the well-being of the ocean.

The discussion below is based on my observations drawn from different areas of the operations on a luxury yacht.

Practices and Procedures

For many years it has been widely demonstrated throughout the industry that the most important goal for crew working on board a luxury superyacht is to provide a luxury service for the owners and guests, and the most valued commodity in order to provide such a luxurious service, was time. There never seemed to be enough hours in the day for the crew on board to carry out every task necessary to maintain such high-quality standard. Therefore, it is observed that over time specific practices and systems began to form and develop to minimize time spent on small daily tasks and optimize time spent creating and providing the most memorable and luxurious experience for the guests that were on board at any given time.

These systems include the heavy reliance on single-use plastics such as plastic zip lock bags for organization, which are used once to store something briefly and then thrown away; the common use of plastic cotton buds to detail and clean tight spaces in the interior of the yacht (which could be on average around 3–5 cotton buds per square metre); throwing out and replacing guest's plastic water bottles even if only a few sips had been taken; the heavy reliance on single-use aluminium coffee pods for crew and guests; the constant washing, drying and ironing of guest clothing every day even if they had only worn an item for one or two hours; emptying and replacing guest rubbish plastic bin liners several times a day even if they only had one or two items in them and throwing away copious amounts of perfectly good food that was only prepared that same day or the day prior in order to save storage space.

It is estimated that in one season an average 45 metre superyacht will go through 5400 plastic water bottles, 6000 aluminium coffee pods, 3600 paper napkins and 6000 plastic straws (www.etyc.fr, 2018).

It has been observed that over time through gained knowledge and experience passed down through generations of yacht crew, these unsustainable systems and practices have become a common theme on board each yacht. However, for some time they were most commonly passed down between crew working on board the same vessel. The yachting industry was a very private industry and there seemed to be a lot of disconnect between different yachts and never much knowledge of information exchanged between vessels, except for maybe between Captains about sailing regions. Therefore, it seemed that once a system was in place on board a particular yacht, it did not seem to change much and this made it very difficult to implement any change on board or make sustainable improvements to these long-standing systems.

There was a general understanding that time spent between yacht crew from different vessels was instead used for relaxing, sharing stories and having a few beverages at the local marina bar. Unfortunately, although these long-standing systems and practices saved time and provided a luxurious service, they were not always very environmentally friendly. The use of vast amounts of chemical-based cleaning products to keep everything on the boat looking immaculate, relying heavily on single-use plastics to save on time and efficiency, generating copious amounts of food waste to provide fresh, high-quality meals and burning large quantities of fuel in order to sail around to diverse corners of the globe, all of these practices contributed to a fairly large eco footprint and many of them have a direct effect on the well-being of the ocean and its marine life.

Cultural Systems

It has been widely demonstrated that when it came to systems on board a superyacht, there were not only practical systems commonly carried out on each vessel, but also many cultural systems that had been widely adopted throughout the industry. For example, lower ranking or newly hired crew members not feeling able to speak up or have an opinion on most matters on board, Captains having the overruling say on all procedures and practices on board even if they did not entirely understand many of them (such as matters around sustainability and environmental awareness) and an overall common 'throw away mentality' that was carried throughout the industry, in which everything on board a luxury superyacht must look immaculate and 'fresh', therefore many items on board were very rarely reused or recycled but were instead thrown away and replaced on a regular basis.

It is also important to note that luxury yachting has evolved from commercial shipping, which has been around for thousands of years, dating back to the bronze ages where of course topics such as climate change and sustainability were not in existence (www.study.com/the-bronze-age-ships-and-rigging—Muscato, 2021). Needless to say, this strongly suggests that the sustainability culture adopted within the luxury superyacht industry did not exactly start out very 'forward thinking' or 'progressive'. It is evidenced that many of the archaic values that started out in the industry when it first began did not accommodate or consider sustainability or the well-being of the ocean. In fact, the first luxury motor yacht was launched in 1908 (which was Charles Henry Fletcher's 34 metre MY Jemima F. III.) (www.nytimes.com, 2018) and it wasn't until 65 years later in 1973 when MARPOL—The International Convention for the Prevention of Pollution from Ships was formed (www.britannica.com, 2020), that there were any governing bodies or laws in place around the prevention of pollution from a ship. The main goal had always been to sail a boat from one destination to another and to provide a luxurious service while doing so. In fact, speaking of sustainability and environmental awareness was almost a 'taboo' subject for a very long time. There have been countless stories shared by yacht crew who have tried to implement change on board but were met with great resistance from fellow crew members (often higher ranking and longer standing members). It has been commonly expressed that many crew working within the industry believed the phrase 'sustainable-yachting' to be a complete oxymoron and that the eco footprint of the industry is so large already that there was no point trying to make even a small change.

Collective Mindset

It is widely recognized that guests come on board a luxury superyacht not only to receive luxury service and hospitality but also to explore and enjoy the beauty, diversity and the very nature of the ocean. Very few tourism industries can provide this unique experience. But for some time, evidence strongly suggests that there was a great disconnect between the yachting industry and its intrinsic connection to the well-being of the ocean. There seemed to be a common lack of knowledge and understanding that the health and abundance of the ocean is the very reason why luxury yachting tourism exists in the first place. Although the crew were intent on providing a luxury service on board at any cost, they were not always consciously aware that the very nature of our industry relies heavily on the well-being of the ocean. Many systems and procedures on board were in fact directly harming the very place they sailed upon for pleasure each and every day.

When media coverage of climate change, ocean pollution and the plastic waste crisis began to peak around 2016 due to the release of independent analyses by NASA and the National Oceanic and Atmospheric Administration (NOAA) which announced that 'Earth's 2015 surface temperatures were the warmest year since modern record keeping began in 1880' (www.nasa.gov.com, 2016) this heightened interest around climate change along with the release of popular documentaries such as Leonardo DiCaprio's 'Before the flood'. As a result, it was widely documented that a vast number of the world's population began to look at making positive changes to the way they were carrying out daily practices, and started to think twice about how they might be contributing to these problems through petrol consumption, using single-use plastics and methods of chemical waste disposal.

In contrast to this, it has been observed that the yachting industry, at first, did not seem to change much. The majority of the luxury yacht industry remained somehow detached and disconnected from the rest of the world's need to take immediate action. The industry lived in its own bubble where the most important goal was to provide a luxury service, at any cost. Not to mention that in order to make any drastic changes it would mean breaking away from the trusted systems that had been heavily relied on for many years. It seemed that although many people working within the yachting industry understood the need to take action against climate change and were doing so in their own personal lives back home, it was difficult to carry these over to life on board a yacht which had a very unique environment and specific set of systems in place for daily operations. There was also very little access to information on which types of sustainable improvements could be made on board. Additionally, there was the fear and uncertainty around change and the unknown, for example, the fear of using cleaning products that had not been tried and trusted for years and the fear of going against the grain and stepping out of the mold that had been created over time. Many crew even feared being ridiculed and possibly losing their jobs for speaking up about such changes.

The Influence of Reality Television, Social Media and Online Shopping on the Yachting Industry and its Culture

Reality Television

Evidence strongly suggests that the change in culture and mindset first began to happen around the time when the US reality T.V. show 'Below Deck' began to gain worldwide recognition around 2017/18. This brought the daily lives of what went on on board a luxury superyacht to the world stage. Suddenly the walls around the yachting industry had been brought down. Up until this point, the inner workings of the industry were not widely known. This was one of the first factors that broke down the barriers of detachment between the yachting industry and the rest of the world.

Social Media

The next factor to break down barriers of detachment was the growing popularization of social media around 2018. Instagram and Facebook accounts created by yacht crew started appearing frequently in the virtual world. These platforms were public outlets where crew members could share raw and realistic insights into the daily life of a superyacht crew member. They were also a place for yacht crew to share knowledge and insight between one another, recommending products, methods and systems, as well as sharing best practice on how to best implement them. There were also many websites, support groups and networks created to help crew in dealing with many common issues that they were facing on board, such as implementing sustainable practices and facing resistance to these changes.

Suddenly, the privacy surrounding the industry and individual yachts had also been removed and the disconnect began to diminish. It was the start of a new era. There are now hundreds if not thousands of yachting-related blogs, websites and social media pages available online. When The Green Stew blog and Instagram page was first released back in 2018, it was one of the first sustainable yachting-minded platforms of its kind, now in 2021 there are now many eco yachting pages available online and more appearing regularly. A few examples are Seastainable Yachting which was created by a stewardess and is a platform offering a range of eco products and resources available to yacht crew (www.seastainableyachting.com, 2021), Ethical Yacht Wear which is also run by a stewardess who has created eco-friendly and ethical crew uniforms as well running many eco-conscious initiatives such as beach clean-ups (www.ethicalyachtwear.com, 2021) and The Conscious Stew who is a stewardess that shares her eco tips and tricks on Instagram, for example how to make your own cleaning products on board and many other sustainable practices.

Online Shopping

Increased access to the world wide web and online shopping created the next great shift in the sustainability movement. For many years limited availability and access to viable eco alternatives had been a huge hurdle the industry has faced. Tried and tested cleaning products that perform well and are safe to use on board a luxury yacht with many delicate surfaces can be hard to source and when it comes to travelling around the globe to various locations there is often very little access to familiar products. This meant that many crew would have a few preferred products that they would trust and that were easily accessible so they would continuously use them. In the last few years, it has been observed that the market for sustainable products within the industry grow exponentially. For example, Ecostore which is an eco-friendly cleaning and body product range from New Zealand that has been around for over 28 years, has now opened up a marine division, which is run by an ex superyacht captain. Ecostore Ocean now supplies products on a large scale to superyachts all around the world. They have also brought out a readily biodegradable, ocean-friendly boat wash and provide large bulk quantities of eco cleaning products to enable yachts to set up their own refill stations on board (www.ecostoreocean.com, 2021). Nowadays there is a plethora of information online and shared knowledge between crew on social media platforms and worldwide access to products, which has made the switch to eco alternatives a much easier task.

There are now many companies that produce and offer yacht-specific products that are easily accessible worldwide. Yachts can now order products online and have them delivered to almost every part of the world. Much less time is spent now on trying and testing out different products on board because there are reviews, guidance and suggestions from fellow crew members which are readily accessible on the Internet.

A powerful role the luxury yachting industry has to play as a large, global industry is the ability to 'vote with its dollar'. Luxury superyachts make large purchases every day, from the food and the cleaning products we use, to the guest amenities and even the uniforms worn by crew. Now that the industry is beginning to turn towards using sustainable and ethical products, this is a huge vote in the right direction for the future of the planet and it supports the people and the companies that create products which align with a greener future for all. For example, sustainable items such as bamboo toothbrushes are becoming much more common onboard yachts (particularly for guest toiletries).

If every yacht made the simple switch from plastic toothbrushes to toothbrushes made of bamboo, which is a much more sustainable material, this would not only have a significant reduction in the plastic footprint of the industry but also the carbon footprint, because bamboo produces 35% more oxygen than a tree of equivalent mass and it can also absorb as much as 12 tonnes of carbon dioxide per hectare per year (www.greenmatters.com, 2021). Another way in which each yacht can make a vote in the right direction with large-scale purchases would be supporting companies that are donating portions of their proceeds to environmental conservation. For example, Ethical Yacht Wear is a crew uniform company that donates 10% of each sale to ocean clean-up and conservation. If every yacht made the switch to supporting products and companies such as these, it can have a profoundly positive impact for the environment on a much broader, global scale.

The Growth of Sustainable Practices on Board a Superyacht

Plastic Waste

It has been observed that reducing single-use plastic on board is now becoming a huge target within the industry. Single-use water bottles are one of the most commonly found plastic waste items on board a superyacht. This is primarily due to the fact that many older vessels do not have sufficient water filtration systems installed and therefore crew are unable to or are hesitant about drinking the tap water on board. It is estimated that for every 10,000 crew employed, 3.2 million plastic bottles are consumed each year (www.clearoceanpact.org, 2020), not to mention the amount of energy it takes to transport these heavy bottles to and from the yacht and the fuel consumption essential for storing and transporting them on board. Another form of plastic waste that tends to go under the radar is micro-plastics that are released into oceans from laundering synthetic materials on board. Every 100 wash cycles discharges up to 100 g of micro-plastic into the ocean (www.clearoceanpact.org, 2020). These millions of micro-plastic particles come from crew uniform and bedding that are washed regularly and the wastewater is then flushed directly down the vessel's drains and into the ocean.

There are now many ways in which yachts are reducing their plastic waste, such as installing water filtration systems on board and providing reusable drink bottles for crew and guests, switching to biodegradable alternatives for single-use items such as straws, cotton buds, toothbrushes, bin liner bags, setting up refill stations on board for cleaning and toiletry products so the same bottles can be refilled and reused countless times. There are also now companies and organizations dedicated to targeting and reducing plastic waste on board superyachts such as The Clear Ocean Pact (www.clearoceanpact.com, 2020), who have created a pact that yachts and crew members can sign to pledge their efforts to reduce plastic waste on board. Currently there are 130 yachts who have already signed up to this initiative. They also provide knowledge and guidance on how to make these changes and which specialized products can be used on board to do so. In addition, micro-plastic filters have become available on the market that are being installed into yacht laundry pipes to catch and prevent micro-plastics flushing into the ocean and crew uniform companies are also bringing out eco-friendly clothing alternatives made from natural fibres such as bamboo, hemp and cotton instead of synthetics. For example, Ethical Yacht Wear is a company that now offer crew uniform made from ethically sourced 100% organic cotton (www.ethicalyachtwear.com, 2021).

Chemical Waste

Chemical waste generated by the vast number of toiletry and cleaning products used on board is another contribution to the environmental footprint of the yachting industry. To maintain a high level of cleanliness on board a vessel, various cleaning procedures are in place for each department, all of which use a wide variety of different products:

laundry detergents, glass cleaners, wood polish, multipurpose sprays, teak scrub and boat wash, to name a few. Additionally, there is a range of different toiletry and body products used on board by crew and guests. These chemicals are used on board in large quantities every day and are all washed down the drains and eventually flushed directly into the marine environment below. Many common cleaning and toiletry products contain a wide variety of chemicals that can be extremely detrimental to the ocean and its inhabitants.

A prime example of commonly found toxic ingredients that can cause direct environmental damage are oxybenzone and octinoxate which are found in common sunscreen. These chemicals have now been proven to cause harm to coral reefs and marine life (www.ncbi.nlm.nih.gov, 2008). The chemical triclosan which is found in many cleaning and toiletry products labelled 'antibacterial' has been proven to have toxic effects on marine algae (www.sciencedirect.com, 2014), it has also been strongly suggested that exposure to the chemical group called phthalates found in many common body washes, shampoos and fragranced body products can cause endocrine disruption in mammals, and these chemicals were detected in the urine of bottle nose dolphins in 2018 (www.agupubs.onlinelibary.com, 2018).

Maritime Pollution Regulations (MARPOL), and more specifically Annex I—Regulations for the prevention of pollution by oil and Annex II—Regulations for the control of pollution by noxious liquid substances in bulk (www.lr.org, 2021), govern avoidance of direct release of toxic chemicals into the ocean, particularly oil, petrol and even exterior boat wash in bulk quantities, however, these are only very loose guidelines to monitor the release of toxins found in common cleaning and toiletry products used on board by crew and guests, such as laundry powders, cleaning detergents, shampoos and body washes. Fortunately, nowadays hundreds of eco-friendly, toxic-free products are available on the market, many of which are specific to the luxury yachting industry and the unique materials encountered on board, which makes it much easier to switch to toxic-free alternatives and reduce this form of chemical pollution. Tighter restrictions around these types of chemicals being used on board would ensure a more profound impact on the preservation of the ocean.

Recycling

Recycling on a yacht has always been rather complex compared to land-based systems. Many crew members come from different countries and cultures where they may have never recycled before. As yachts are often sailing between countries, this means that shore-based recycling systems are regularly changing. For instance, France has a different recycling system from Italy, and Italy has a very different recycling system from that of the United States or New Zealand. This can create a lot of confusion and lack of enthusiasm and trust in the systems that are already in place. There are also many ports and marinas that do not provide recycling bins or have recycling systems. One of them, surprisingly, is a popular Marina in Fort Lauderdale. It can be extremely disheartening when the crew have been vigilant with separating recycling on board only to find that it ends up being thrown into one general waste bin onshore that goes straight into a landfill.

Despite strict MARPOL regulations around recycling on board which can be found in 'Annex V Prevention of pollution by garbage on ships' (www.lr.org, 2021), many yachts still do not follow them. Unfortunately, there are no regulations around Marinas and Ports having to provide sufficient recycling bins on the dock. There is clearly a need for more regulations around this. Additionally, if the region has the facilities and capability to recycle, having bins available in every Port and Marina for vessels to be able to responsibly dispose of their recycling would be highly advantageous. Fortunately, there are already people campaigning to change this, for example, The Clean Superyacht Marina Campaign was created by people working within the superyacht industry who have noticed the demand from crew for more recycling facilities in Marinas and Ports worldwide. 'It has been set up to investigate current facilities in worldwide superyacht marinas and to hopefully start a conversation with marina operators to improve the current facilities and contracts' (www.cleansuperyachtmarina.com, 2020).

Fuel Consumption and Air Pollution

Fuel consumption and air pollution have always been primary contributors to the yachting industry's large eco footprint. Superyachts burn many tonnes of fuel every year in order to sail around the globe and generate enough power to run all of the systems on board. For example, an average sized 71-metre yacht burns approximately 500 litres of diesel per hour (www.towergateinsurance.co.uk). Fortunately, there are now many new innovations in the technology around engine development and naval architecture that are helping to lighten the footprint. For example, the use of bio-diesel is becoming more common, many boat builders and designers are turning to hybrid propulsion, hydrogen engines and fully electric engines. 'Feadship's 83-meter Savannah is the first superyacht to use an eco-friendly hybrid propulsion platform encompassing propellers and azimuthing thrusters for power, electric motors and diesel engines for drive, and gensets and batteries for energy storage' (www.luxurylondon.co.uk, 2021).

The ability to harvest and use renewable energy generated by wind and solar power is becoming more favourable as well as installing filtration systems for exhausts to lower levels of nitrogen oxide released into the atmosphere. In 2017, Oceano launched the largest ever DynaRig yacht 106.7-metre Black Pearl. With its regenerative technologies, the three-masted ship can cross the Atlantic Ocean without using a single drop of fossil fuel. Its system relies on the use of the yacht's speed under sail to generate electricity with a variable pitch propeller. 'German yacht builder Lurssen has developed a filtration system that lowers levels of nitrogen oxide (NOx), the gas responsible for forming smog and acid rain, which are both detrimental to the tropospheric ozone layer' (www.luxurylondon.co.uk, 2021). Naval architects 'Lateral Engineering' have also released the concept for a hydrogen-powered superyacht which means it would only emit water. 'Project AQUA utilizes 100% alternative fuels to present a technically audacious view of a zero emissions future. AQUA has been developed around a hydrogen-electric propulsion and energy system architecture' (www.lateral.engineering.com, 2019).

Education

Nelson Mandala once said, 'Education is the most powerful weapon you can use to change the world'. Unfortunately, education around sustainability has not yet been on the priorities list for crew training on board a superyacht. The basic training and certification necessary to work on board a superyacht at entry level it based around STCW Basic Safety Training which encompasses 5 elements which are Personal Survival Techniques (A-VI/1-1), Fire Fighting and Fire Prevention (A-VI/1-2), Elementary First Aid (A-VI/1-3), Personal Safety and Social Responsibility (A-VI/6-1) and Proficiency in Security Awareness (A-VI/6-1). Food safety and hygiene has also recently become mandatory for any crew member that is handling food on board.

With the way the industry is heading with its sustainable endeavours it is hoped that the next mandatory course necessary for working on board a superyacht for crew will be based around sustainability and environmental awareness. There are already several courses and certifications around these topics that have become available online over the last few years for yacht crew. An example of training already on offer for crews is ETYC (Environmental Yacht Crew Training). ETYC is the first company offering training, audit and consulting to yacht crew to implement best practice to protect the environment. Inspired by the ISO14001 standard, ETYC is offering solutions aiming at implementing Environmental Management Systems (EMS) and developing proposal to improve environmental performance on board yachts (www.etyc.fr, 2021).

Sustainable Tourism

Over the past ten years, the luxury yachting industry has evolved in its eco endeavours in many ways. Evidence suggests that a powerful drive for this change has been consumer demand. We are now commonly seeing how consumers are choosing sustainable and ethical products for their everyday purchases, as well as when choosing a holiday. People are always going to seek holidays and excursions that are in alignment with their values, and for more and more people these days those values are to uphold and protect the environment. In 2017 according to a survey conducted by booking.com, one in three tourists intended to choose options that are more respectful towards the planet (www.uoc.edu.com—Font, 2017). Soledad Morales, professor at the UOC Faculty of Economics and Business, explains 'As we have become aware of the negative effects of certain forms of tourism, concepts such as sustainability and responsibility have gained more importance' (www.uoc.edu.com—Font, 2017).

This also suggests that when luxury yachting clienteles are choosing which yacht they are going to buy or charter, it seems the larger majority are now more likely to look at a yacht that has a lesser ecological footprint and carries out sustainable practices. 'New generations of superyacht buyers are increasingly focused on solutions that protect the environment and when approaching their own yacht project are attracted towards solutions that can address this issue' said Daniele Bottino from the American Bureau of Shipping (ABS) (www.superyachtnews.com, 2020). It is commonly understood that consumer demand has been a huge driving force in the progression of sustainability in the industry which has experienced an increased interest in both diesel-electric and hybrid systems, according to Andrew LeBuhn, a top yacht broker at Camper and Nicholsons (www.worth.com, 2019).

As part of the tourism industry, luxury yachting has the opportunity to give back and to lead by example for many other large industries around the world by changing the ways that it operates. It is in a unique position as a global industry that has access to many remote corners of the globe that other industries do not. This can be used as a means to not only spread awareness and inspire communities worldwide to make a difference but also as a means of aiding and supporting local communities in need. An example of this is in the humanitarian and conservation work done by Yacht Aid Global which is a non-profit organization which encourages and supports superyachts in providing humanitarian aid, conservation and disaster response to the remote regions that they sail to. According to Mark Drewelow, Founder of Yacht Aid Global, their work 'encompasses global programs, logistics management and consultation for superyachts which want to contribute to the world around them.

We facilitate volunteering by yachts and crew as they work on humanitarian and conservation initiatives in the communities they visit' (www.yachtaidglobal.com, 2016).

Another smaller example of inspiring change is when a crew member visits a local bar or restaurant in a remote village and asks the waiter for no plastic straw with their drink and then explains the reasons why, this can create a ripple effect, causing waves of discussion around the use of single-use plastics in the local community. Thus, local bars can make a decision to stop using plastic straws. We have seen this wave of change occur in many top tourism destinations such as Thailand and the Philippines. This can spark interest and inspire possible change or at least discussions within the community around these topics. History shows that leading by example can be a powerful way to create positive change.

The Future of Sustainable Luxury Yachting

The future for sustainability in the luxury yachting industry is looking very promising. In a short space of time the industry has already made some significant progress. There is now an abundance of initiatives and organizations promoting and driving sustainable practices and programmes throughout the industry, such as The Water Revolution (www.waterrevolutionfoundation.org), The Superyacht Eco Association (SEA) Index (www.superyachtecoindex.com), Eat Less Plastic (www.eatlessplastic.com), The Clear Ocean Pact (www.clearoceanpact.com), Seastainable Yachting (www.seastainableyachting.com), The Conscious Yachting Club (www.consciousyachting.club), Clean Sailors (www.cleansailors.com), Yacht Aid Global (www.yachtaidglobal.org) to name a few.

Many superyacht industry leaders are now beginning to work together to find sustainable solutions. For example, The Water Revolution Foundation, a non-profit organization which includes several major shipbuilders (including Benetti, Feadship, Heesen and Lurssen among others), design companies, technology companies and many others associated with the yachting industry, pursues a goal to create a science-driven approach to yachting and neutralize the industry's ecological footprint. The organization provides much needed guidance, assessment and tools for companies seeking to make sustainable change (www.waterrevolutionfoundation.org, 2020). The use of recycled, reclaimed and renewable materials is becoming more common in the yacht construction process, and there are already many yachts that have sea bins installed on board their vessels which collect ocean waste while they are underway. Many large yacht manufacturers and shipyards are even taking pledges to reduce their greenhouse gas emissions for the upcoming years and committing to prioritizing ocean conservation. As an example, Oceanco, a Dutch shipyard, hopes to use 100% renewable energy for its electricity usage and be 100% waste-free or circular across the entire supply chain by 2030. They have already begun installing renewable energy systems in their facilities and its Alblasserdam facility currently generates 250,000 kWh per year of renewable energy. The end goal is to be one of the world's most sustainable yacht builders (www.luxurytribune.com, 2020).

The superyacht industry is in a prime position to lead the way in innovation and technology around sustainability, supported by high net worth superyacht owners, highly skilled people behind yacht production, and crew. As noted by The Water Revolution, 'The superyacht industry is the perfect breeding ground to trial, test and perfect new sustainable solutions. This is thanks to our affluent clientele who have the resources to drive sustainability, and the many skilled craftsmen, engineers and architects who are not afraid of pushing the boundaries. Now is our opportunity to build a sustainable yachting industry that is future-proof and attractive to the next generation of clients' (www.waterevolutionfoundation.org, 2021).

Conclusion

The observations of the progression of sustainability within the yachting industry have been shared from a practitioner's perspective. Evidence from practice demonstrates significant disconnect on many levels which has played an important part in delaying the initial progression towards environmental consciousness in luxury yachting. However, notable progression and evolution can now be evidenced in many aspects, such as culture change, technological advancement, interconnectedness through social media and the world wide web. There is now a great sense of purpose and drive towards sustainability agenda that is fuelled by crew members, yachting organizations and industry leaders. The industry is moving away from the old traditions and making shifts towards creating and implementing many new systems which better accommodate and preserve the well-being of the ocean and the environment. There are signs that the culture throughout the industry and on board is beginning to change with fresh perspective shared by yacht owners and crew.

Communication via social media has influenced interconnectedness of crew from different yachts which has allowed the sustainable yachting movement to gain great momentum. Personal practitioner's experience as a Stewardess and ventures of The Green Stew have evidenced that many crew now actively seek sustainable alternatives to the once 'set in stone' systems, practices and products that were in place. Many yachts already use less plastic, less toxic chemicals and more reusable products while they continue to successfully provide a high-end luxury service. The proof of this evolution can be observed through the increase of sustainable yachts that have been built and launched over the past five years, and in the fact that there have been many amendments made to the older marine pollution laws since 2016 (www.lr.org, 2021). From my experience it is safe to say that although providing a luxury service is still top priority, it is becoming apparent that luxury and sustainability can work hand in hand and it doesn't have to be binary. With the powerful combination of eco-conscious people, education and new technology there are excellent possibilities for sustainable practices in the luxury yacht industry.

References

AGU Advancing Earth and Space Science—Geo Health. (2018, September 5). *Urinary phthalate metabolites in common bottlenose dolphins (Tursiops truncatus) from Sarasota Bay, FL, USA.* Retrieved April 27, 2021, from https://agupubs.onlinelibrary.wiley.com/doi/10.1029/2018GH000146

Britannica. (2020). *Regulation, international conventions.* Retrieved April 14, 2020, from https://www.britannica.com/technology/ship/Regulation#ref528333

Clean Sailors. (2020). *The new standard in sailing: Clean Sailors, for cleaner seas.* Retrieved June 19, 2021, from https://www.cleansailors.com/

Clear Ocean Pact. (2020, December 10). *Home page.* Retrieved January 15, 2021, from https://clearoceanpact.org/

Conscious Yachting Club. (2020). *Fast track access to a culture of sustainable development and ESG practices for the SuperYacht Industry.* Retrieved June 19, 2021, from http://consciousyachting.club/

Eat Less Plastic. (2019). Retrieved April 10, 2021, from https://www.eatlessplastic.com/

Ecostore Ocean. (2021). Retrieved June 10, 2021, from https://ecostoreocean.com/

Ethical Yacht Wear. (2021). Retrieved April 7, 2021, from https://ethicalyachtwear.com

ETYC—Environmental Yacht Crew Training. (2018, October 19). *About us.* Retrieved April 27, 2021, from https://www.etyc.fr/en/

Font, R. (2017). *Sustainable tourism is becoming increasingly popular.* Retrieved April 20, 2021, from https://www.uoc.edu/portal/en/news/actualitat/2017/024-sustainable-tourism.html

Green Matters. (2021, January 19). *What makes bamboo sustainable?* Retrieved April 27, 2021, from https://www.greenmatters.com/p/why-is-bamboo-sustainable

Lateral Naval Architects. (2019). *Projects.* Retrieved April 7, 2021, from https://www.lateral.engineering/projects/aqua/

Lloyd's Register. (2021) *MARPOL International convention for the prevention of pollution from ships.* Retrieved April 7, 2021, from https://www.lr.org/en/marpol-international-convention-for-the-prevention-of-pollution/

Luxury London. (2021, February 11). *Turning the tide: The sustainable future of superyachts.* Retrieved January 6, 2021, from https://luxurylondon.co.uk/drive/yachts-and-jets/sustainable-superyachts

Luxury Tribune. (2020, December 26). *The rise of sustainable yachting.* Retrieved April 26, 2021, from https://www.luxurytribune.com/en/the-rise-of-sustainable-yachting

Muscato, C. (2021). *Bronze age ships and rigging.* Retrieved April 14, 2021, from https://study.com/academy/lesson/bronze-age-ships-rigging.html

NASA. (2016, January 21). *NASA, NOAA analyses reveal record-shattering global warm temperatures in 2015.* Retrieved January 6, 2021, from https://www.nasa.gov/press-release/nasa-noaa-analyses-reveal-record-shattering-global-warm-temperatures-in-2015

National Centre for Biotechnology Information. (2008, January 3). *Environmental Health Perspectives.* Retrieved April 27, 2021, from https://www.ncbi.nlm.nih.gov/pmc/articles/PMC2291018/

NY Times. (2018). *Queen of motor yachts; The Jemima F III., Largest in the world, Launched in Harlem.* Retrieved April 14, 2021, from https://www.nytimes.com/1908/07/17/archives/queen-of-motor-yachts-the-jemima-f-ill-largest-in-the-world.html

Science Direct. (2014, July 15). *Marine pollution bulletin.* Retrieved April 27, 2021, from https://www.sciencedirect.com/science/article/abs/pii/S0025326X14002884

Sea Index, Superyacht Eco Association. (2021). *Homepage.* Retrieved April 7, 2021, from https://superyachtecoindex.com/

Seastainableyachting. (2021). Retrieved June 19, 2021, from https://www.seastainableyachting.com/

Superyacht News. (2020, June 4). *Is the Superyacht sector ready to switch over to hybrid power?* Retrieved April 27, 2021, from https://www.superyachtnews.com/technology/is-the-superyacht-sector-ready-to-switch-over-to-hybrid-power

The Clean Superyacht Marina Campaign. (2020). *Home page.* Retrieved April 27, 2021, from https://www.cleansuperyachtmarina.com/

Water Revolution Foundation. (2020). *About.* Retrieved March 20, 2021, from https://waterrevolutionfoundation.org/about/

Worth. (2019, November 25). *Is Green Yachting the Next Wave?* Retrieved April 27, 2021, from https://www.worth.com/is-green-yachting-the-next-wave/

Yacht Aid Global. (2016, October 14). *About us.* Retrieved April 27, 2021, from https://yachtaidglobal.org/about-us#whatwedo

Yachtworld. (2020, July 11). *Superyachts and sustainability.* Retrieved January 10, 2021, from https://www.yachtworld.com/research/superyachts-and-sustainability/

7

Innovative Galley Management for Sustainability in Luxury Yachting

Thean Labuschagne

Introduction

My career in luxury yachting started after six years of culinary training and working as a chef at land-based restaurants where I was purposefully moving from one section of the kitchen to the other in order to perfect my skills as an all-rounder chef. I have worked with chef Michael Broughton, Annette van den Water, Ruan Pretorius and Hildelee Olivier, to name a few, and practised and perfected my skills at Terroir Restaurant, one of the top ten restaurants in South Africa at the time. With all this knowledge and professional confidence, I stepped into the world of yachting in 2017, where I was very lucky to work with chef William Fiddy, my first head chef on the yachts, and still my head chef today, who taught me most of what I know today concerning the yachting industry. It is not an easy task converting a land-based chef into

T. Labuschagne (✉)
Luxury Yacht Practitioner, Vereeniging, South Africa
e-mail: jahthean@icloud.com

T. Gladkikh et al. (eds.), *Luxury Yachting*,
https://doi.org/10.1007/978-3-030-86406-4_7

one that is of value to a yacht—there is a big difference between the two. I also learnt exceptional skills for a yacht chef from Andrew Murphy, my other major influencer. These skills are related to administration and organization of a professional galley. A luxury yacht chef is expected to anticipate and handle every arising issue as their second nature. He/she must be in control of the galley, stay ahead by thinking in a specific manner and pay attention to each sector of the galley, which is not as easy as it sounds considering that the boat does not always have access to provisions, therefore extensive and careful planning is a must.

I approach my professional responsibilities with sustainability in mind. My interest in the environment and the role I play in sustainability and growth was deeply imbedded in me during my years at University where several professors changed my entire outlook on life by explaining what the world would suffer as a consequence of our ignorance. I continue living by that rule and try to be the best I can when it comes to looking after our beautiful planet. Although I am not the world's most experienced chef, I am certainly an ambassador of its well-being. Therefore, I write about my work as a luxury yacht chef with integrity and truth.

The land-based chef has always had a reason to look for ways to save money and therefore increase the restaurant's profit by minimizing waste and finding innovative ways to use or reuse food in ways never thought possible. This sort of drive has led to a worldwide unwritten chefs' rule, which dictates that any chef with integrity will think twice about wasting food and try to find ways to incorporate things that usually would go to waste into their dishes. Amongst these are having a sort of guideline for portion sizes, calculating the average patron visit during a certain time as to not over provision and not to prepare too much of a certain product which would lead to inevitable waste.

Unfortunately, the same rules do not always apply to most luxury yachts. On most luxury yachts the galley does not work towards a profit and therefore the amount of waste can be shocking. The aim of this chapter is to raise awareness among yacht chefs and guests alike and to open their minds to improve this part of the industry into a more sustainable and less mindless act.

In the next part I will explain the main differences between the yacht chef and the land-based chef and the challenges that both face respectively to manage waste. Inevitably by highlighting these challenges, one can be encouraged to think about solutions.

Yacht Chef versus Land-Based Chef

The land-based chef works mostly on a seasonal menu with a certain amount of guests. The suppliers are always the same and the menu items can be categorized to determine a favourite and least favourite dish, where the least favourite dish would be, for instance, the vegan or gluten-free options on the menu or even the really expensive options such as white truffle, lobster and caviar. The season can also determine the number of guests likely to visit the restaurant and therefore make it easier to determine the amount of preparation, provisions and number of staff needed to operate the restaurant. The supplier will also have seasonal produce which will determine the items on the menu and have a positive impact on budget planning.

The yacht chef, on the other hand, travels from place to place on a regular basis and therefore has to change suppliers regularly. This makes it hard to determine a specific budget and also a menu—this is why numerous yachts simply import their produce at sometimes incredibly high cost. It is also difficult to have a specific menu for guests, as they are allowed to change the menu according to their liking at any time. Most yachts have preference lists for the guests to get a better picture of the produce that should be available on board. But still, it is not easy to get everything that every person likes, just in case they should fancy that while visiting the yacht.

As far as staff meals go, the land-based chef has to provide a meal once per day for their staff and use off cuts of meat available in abundance along with a carb and a vegetable or salad. Each person gets a pre-portioned plate and if they are not happy with the meal, they have to make a plan for themselves, for example packed lunch or takeaways.

The yacht chef has to prepare three meals every day for the crew. Breakfast is usually an egg dish with a few varieties of accompaniments and perhaps a baked dish or a fruit platter along with accompaniments. Lunch has to consist of two proteins, a carb and two salads. Dinner is two proteins, a carb and two hot veg. The meals have to vary so that the crew do not eat the same meal twice in a week. These menus are completely made up by the chef and the crew budget is therefore more predictable as it can be pre-planned a week or two in advance.

As far as waste management is concerned, the land-based chef will have a number of contractors that take care of their specific waste needs in their different ways. This will take place on a weekly basis. This can be calculated to maintain a steady stream of waste, making it easier to manage. The yacht chef, on the other hand, will have to plan waste disposal where possible, depending on whether these services are available where the yacht is positioned. This is not always an easy task when it comes to recycling used oil, excess fat, harsh chemicals, etc. This takes meticulous planning and some chefs do not want to go the extra mile to ensure safe waste disposal.

As for actual cooked food waste, most land-based restaurants have big barrels that are collected once a day or every two days. These barrels are sold to pig farmers and in return the pig farmers sell their products to the restaurant at more reasonable prices (this is just one of many ways to dispose of cooked food). It is hugely disappointing that most of the cooked food waste on yachts either goes overboard or ends up in the bin. MARPOL (International convention for prevention of marine pollution) strictly states that a vessel is only allowed to throw food waste overboard when 12.5 miles from the shoreline.

In a land-based restaurant cleaning products of wide variety can be used and the chemical control regulations are fairly standard—harsh chemicals may be used when disposed of correctly, while yachts only have a certain amount of space to store these types of chemicals (e.g. the ones found in kitchen spec tanks). Yacht galleys have a strict and specific regulation for what exactly can be used on board and this is challenging when taking into consideration how much time is spent on specific cleaning tasks. It is much more time-consuming using environmentally friendly cleaning products, especially in the galley.

Regarding time management, the land-based chef works a specific section, alongside ten or more chefs on a shift, with a specific time schedule depending on the restaurant. The chef will have his own section where he/she is responsible for the stocktake, provision lists, cleaning and maintenance. The weekly, monthly and yearly cleaning and maintenance schedules are divided among the multitude of chefs and can be done in a matter of days. The yacht chef usually works alone or alongside no more than two chefs for very large yachts. The guests' hours are unpredictable and therefore the hours of work for chefs cannot be specifically calculated. On the average guest trip, the yacht chef works about 16–17 hours for an average of about 7–10 days on a trip. The yacht chef also has to see to all the additional cleaning jobs including washing dishes and maintaining the galley. All the provisioning for both crew and guests, the menus and the preparations, are all dependant on this individual as opposed to a group of chefs. This makes time management extremely difficult and as a result some chefs neglect sustainability and waste management protocols in order to achieve guests' expectations.

For a chef to start out his/her career in the land-based kitchen, he/she only needs only their qualifications from culinary school and then advance to learn at each section of the kitchen as they progress in the work situation. Never doing more than one section at a time, the land-based chef learns to focus on a specific cuisine, repeating the same preparation for a few months at a time until the menu changes. The advantage of this is that one can perfect a certain skill before moving on to the next section. Chefs on a yacht have to already have mastered all the sections as they are required to do everything from baking to grilling, plating to washing dishes. Yacht cuisine is based on where the yacht finds itself in the world or it depends on the theme of the guests' meal. Therefore, the yacht chef has to have a broad knowledge of cuisines and the skills to perfect every section of the kitchen by themselves.

The above mentioned are a few of the major differences that I myself have experienced when comparing the two industries with one another. Some may differ, but in a broad spectrum of the two worlds, these differences will always occur. The challenges in general, are far greater for the yacht chef, but the capital behind the yacht chef is far greater than that of a land-based chef. It is therefore very important, especially for the new generation of yacht chefs, to respect and appreciate the means they have to get by. They have to realize the tremendous responsibility one has towards nature to maintain sustainability, obey the rules set in place to prevent pollution and to do everything necessary to reduce waste or to reuse it in ways that will reduce waste. A chef with integrity will not just use the easiest and quickest way of getting things done but be thoughtful about everything from produce and its carbon footprint to reusing things that would normally go to waste in innovative ways—to reincorporate these potential waste products and give them meaning and purpose in the Galley.

Below I shall explore helpful tips, innovative ways of reusing scraps to make them useful on the plate, discuss carbon footprint and the reduction thereof and how to make it rewarding to do all these things in a manner that is acceptable to a chef with integrity. The real challenge lies in getting people to understand how important it is for every chef to do his/her little part for the greater good.

Helpful Hints and Tips

I shall now focus on the yacht chef only. I shall discuss the problems that a yacht chef faces from both the galley and a traveller's perspective and suggest a few possible solutions for these problems. The advantage of having prospects is that they can always lead to the change for the better. Thus, the more we concentrate on the problems we are facing right now, the better solutions we can come up with in the future. Luxury yachting is an ever changing industry in an ever changing world around us, therefore it is important to appreciate and be prepared for uncertainty (Weick, 2015) to keep up with the latest trends and work on solutions according to the latest developments.

Edd McCracken (2018) claims that in a matter of years water will become the world's most valuable commodity. For this reason, every chef needs to be respectful towards water usage and try their best to conserve this ever depleting resource. There are numerous ways that a chef can save water. Most galleys have sanitizers that run on a quick 90-second cycle. In order to use them properly, one needs to rid any utensils of debris and simply place them in the sanitizer. The cycle exposes the utensils to a temperature of 90 degrees Celsius for 90 seconds and therefore rids the utensils of the most harmful bacteria that can cause food-born illnesses. The cycle uses a lot of water, but it is a necessity to make use of this in order to maintain a safe environment. One can save a lot of water by having a sink full of soapy water and rinsing the utensils, pots or pans in the sink rather than using flowing water every time a utensil needs to be rinsed. One can also make sure to fill the sanitizer up before running a cycle and therefore reduce the amount of water used.

A system for waste needs to be put in place and carefully monitored. Separating recyclables, cooked food, raw meat, raw vegetables and raw fruit is essential when it comes to reusing waste or finding ways to make waste work rather than just throwing it away. In our Galley we keep raw, hard vegetable peels and offcuts, then dehydrate them (in other words, dry out) on a low temperature. They are then used to make stocks and broths, or as a base for some sauces. In this way we save a lot of waste and make it work for us at no extra cost. In the same way, all meat offcuts can be used to make stocks, sauces or broths by simply adding a few ingredients and extracting the flavours. Although this is not always possible, some ports do have pig farmers who will collect any cooked food from the yacht and sometimes even provide barrels to collect the food in. It is, however, the chef's responsibility to not prepare too much food by exercising portion control, not prepare food too far in advance so the food may spoil before serving, check the dates of produce which expire soon and use it up before those that expire on a later date (first in, first out) and know the guests and crew to avoid preparing food that they are not fond of.

Other methods of reusing food include most dried ingredients that are used to flavour sauces or soups. For instance, when I make a sauce with expensive dried porcini mushrooms, after straining the sauce, the mushrooms are then used to make a dark velvety puree by just adding a few ingredients and then blending them to the right consistency by adding cream. This is just one of many examples of dried produce that can be remade into something different to be featured on another plate. Other examples include cake offcuts that can be dehydrated and put into a food processor to create a crumble. This can feature on dessert plates by placing an ice cream on it, so it doesn't slip off the plate. Stale bread can be processed into a crumb that can be used as a coating or used in pestos or as a thickening agent for soups. Tomato offcuts can be used along with egg whites to clarify stocks or consommés. Some chefs burn vegetable peels and process them into an ash which is used to clarify stocks or make a black dust for colouring or coating. Fruit offcuts are mostly frozen and can be used in smoothies or coulees. Potato peels are fried and flavoured and served as a bar snack. If it is not possible to find the time to reuse these scraps, most places, e.g. San Fransisco, have established recycling food waste to produce compost heaps, which in return is used to fertilize the soil for local gardens or for farmers. Used oil can be strained and sold to places that reuse it to make biofuel. Fat offcuts can be rendered and strained and then used to cook the meat or potatoes in for extra flavour. For instance, duck fat is rendered and strained and used to make excellent roasted potatoes. The possibilities are endless and new ways of reusing appear almost on a daily basis.

Carbon Footprint

The chef's responsibility towards nature and reducing the carbon footprint on the environment can be done in a few ways. These include supporting local markets and buying produce that is currently in season. Supporting local farmers' markets allows to cut out transportation of produce. Additionally, by supporting local businesses, the big corporations that leave a mark on the environment are not supported as often. It is the chef's responsibility to create menus around the seasonal availability of produce. There are shocking stories of, for instance, the avocado trade with respect to import and export. According to the Netflix programme Rotten (Netflix, 2020), gangs run the trade of avocados in places like Chile and use all the water for their crops while people around them are starving and being left with no water for days or sometimes weeks on end. They also charge extortionate prices for water to the locals where most of them cannot afford it. It is also, therefore, a great responsibility of the chef to know where their produce originates from and the circumstances in which these products have been exported.

When it comes to sustainable fishing, most private fishermen selling at the docks or markets are usually the safest option in comparison to larger corporations selling bulk. The local produce is also fresh and more often than not larger in size and better in taste. Large fishing vessels have a massively negative impact on the environment and do not usually fish for a specific species. They use large nets which capture undersized and sometimes endangered species that cannot be sold. These species are sometimes killed by strangulation in the nets that are supposed to catch the fish they can process and sell. Commercial fishing can also damage the sea floor and underwater habitat (Hill, 2021). It is also the chef's responsibility to know which species are caught at a sustainable rate and which ones face endangerment so as to only buy those that are sustainable. These lists can easily be found on the Internet and are more often than not region-specific.

When it comes to cleaning products there are a large variety of options which are easily available nowadays that are environmentally friendly. These come at a cost but they are specifically designed with the environment in mind. It is therefore the chef's responsibility to use them in minimal amounts to avoid waste and also to support these companies rather than those that do not have the environment in mind. These cleaning products are also safe to use when using very fragile sewage treatment plants such as the ones often found on yachts. They do not kill the good bacteria that help restore the water to a safe disposable level.

Maintaining Budgets

It is very difficult to try and maintain a steady budget while travelling around the world. Different currencies and different costs of products needed challenge the chef to an extent where planning becomes crucial.

Most yachts have adequate stores to keep about six months' supply of dried and frozen goods. It is therefore of great importance that the chef gets as much of these products when in places such as Europe or the United States. The reason behind this is that good and affordable produce is harder to find when the yacht is in the Galapagos instead of somewhere in the United States or Europe. Careful planning is necessary to determine how much of a certain product will be used in order to not get too much of it. In this manner, chefs only have to calculate budgets for fresh or perishable products, thus making it easier to maintain a steady budget. When provisioning in such bulk it is extremely difficult to deliver everything on board and organize, so this process may take several days to complete.

While some yachts do not have budgets, it is still important for any chef with integrity not to waste money and spend frivolously.

The Contribution of the Luxury Yachting Industry to a Broader Sustainability Agenda: a Chef's Perspective

My discussion of luxury yachting industry's contribution to the environment is aligned to broader sustainability initiatives and agendas. For example, there are strong links between UN Sustainable Development Goals (SDGs) (www.sdgs.un.org/goals) (SDGs, 2021) and activities on board of luxury yachts which have been discussed above. Below I shall demonstrate further how luxury yachting industry contributes to selected SDGs thus supporting the philosophy of constructive and ethical contribution to society and the planet.

SDG 1: No Poverty

When the yacht crew find themselves in remote and often economically disadvantaged places, they contribute in the effort to fight poverty by purchasing supplies directly from farmers and local businesses. This helps to support local producers raising farmers' profit margins significantly. This essentially means that farmers can sell their product at the market, for the same price as a larger supplier would, this allows them to cut out the suppliers' fee and make more profit for themselves. This in return creates job opportunities in the summer when there is such a great demand for good produce.

SDG 2: Zero Hunger and SDG 17—Partnerships for the Goals

Some yachts form partnerships with charities, and every year around Christmas, yacht crew are encouraged to contribute by donations of any kind. Typically, the interior and galley departments evaluate their past season's stock, identify dry goods which are nearing expiry date or are in abundance, and then donate these food items to charities. Shipyard or harbour administration is able to assist regarding charities with integrity—they will point yachts and crew in the right direction and this is how we usually get in contact with local charity organizations.

SDG 4: Quality Education

The captain of the previous yacht I worked on would allow local young people who were finding it difficult to decide on their future career to come on board over the wintertime and shadow some of the crew members for a certain amount of time in order to inform themselves of the work that actually goes they could do on yachts and to understand interdepartmental structures of the yachting industry. Most yachts also have a budget for crew education. In the last three years, the yachts I have worked on, have sent me to India and Thailand to study on cooking courses and I'm currently enrolled in a Clinical Nutrition course, which is fully paid for by the yacht. Other departments also have the opportunity to make the most of this budget by enrolling in various courses relevant to their professional development. In this way the yachting industry contributes substantially to quality education of their employees.

SDG 8: Decent Work and Economic Growth

The yachting industry as a whole has a significant impact on local businesses from a seasonal point of view. Certain hotspots (Yacht-spots) for summer and winter travel, are known to industry stakeholders. The Mediterranean is the main attraction from late March through to

October, while Florida in the United States and the Caribbean are highly popular from November to February and February to April, respectively.

Local businesses, being aware of this seasonal dynamic, prepare their products and services and also alter their prices for the tourist season. A distinctive drop in price and also the abundance of produce is very noticeable during the quieter/out-of-season months in market places that flourish during the tourist season. Some of the traders focus mainly on the high season where they make the most profit by selling directly to the public and/or yachts, contrary to winter months where specialized ingredients or products are in decreased demand.

Despite the changes towards sustainability which are already taking place in the industry, a lot more work needs to be done with regard to being totally sustainable and environmentally friendly. In my view, the yachting industry, for the most part, is starting to recognize further challenges and changes that need to be made to change for the better. It is disappointing that most of us only see the negative side of the industry that is so often portrayed in the media. To counterbalance this negative perception, this chapter has highlighted the contribution the industry is making already and suggested some ways of enhancing and widening our positive impact on the planet.

Conclusion

It is of the utmost importance for the future generation of yacht chefs and owners alike to familiarize themselves with the challenges we all face in our current time, set reachable goals to improve our situation and create a culture where we are, as much as possible, aware of our surroundings.

It is inevitably our responsibility to look after the Earth that we inhabit and to protect mother nature to the best of our abilities, so the next generation may experience the same wealth this beautiful planet has to offer. It is a mindless and irresponsible act of the intellectual individual to turn a blind eye to the problems we face as humans living on a planet that we are bleeding dry. Working for one of the world's wealthiest industries, we can make a difference by being mindful, vigilant and respectful. I shall leave you with a request: to create a culture where it is fashionable to be mindful, where it is trendy to respect the Earth and where the younger generation can look up to integrity and mindfulness which we can show towards our beautiful planet. For if we do not make the effort to change our ways, it begs the question: Who will?

References

Hill, J. (2021). *Environmental consequences of fishing practices.* Retrieved May 14, 2021, from https://www.environmentalscience.org/environmental-consequences-fishing-practices

McCracken, E. (2018). *Will water be ever worth more than oil?* University of Melbourne. Retrieved April 23, 2021, from https://pursuit.unimelb.edu.au/articles/will-water-ever-be-worth-more-than-oil

Netflix. (2020). *Rotten: The avocado war.* Retrieved May 14, 2021, from https://www.netflix.com/gb/title/80146284

SDGs. (2021). *The 17 goals.* UN. Department of Economic and Social Affairs. Retrieved April 23, 2021, from https://sdgs.un.org/goals

Weick, K. (2015). Ambiguity as grasp: The reworking of sense. *Journal of Contingencies and Crisis Management, 23*(2), 117–123.

8

Riding the Waves of High Sea and Life: a Crew Member's Reflection on Yachting

Katrien van der Merwe

Introduction

Luxury yachting is a world in itself, and working in the industry is a truly unique experience. This chapter specifically aims to explore the working environment of professional yacht crew members and its effect on people who spend their professional life working on board. As it can be found in any professional industry, yachting has its own vocabulary, insider jokes and stereotypes. What differentiates the yachting profession from many others is its transient life. This context underpins the discussion of the industry which has rapidly grown and evolved over the last 20 years and has matured into a much more regulated industry and with more clearly defined career paths.

This chapter discusses four main themes in the context of yacht crew and, to a lesser extent, the point of view of yacht owners in the yachting

K. van der Merwe (✉)
Independent Luxury Yacht Practitioner, Pretoria, South Africa

T. Gladkikh et al. (eds.), *Luxury Yachting*,
https://doi.org/10.1007/978-3-030-86406-4_8

world: exploration, captivity, expectation and reality. These four themes play themselves out in a very small and yet global environment and as such have interesting consequences and manifestations on the crew, as well as guests. They are central to the world of yachting and this chapter will address them as a guide to construct our better understanding of this complex, elusive and under-researched world. It builds on the personal experience of the author and includes personal observations, experiences and conversations that she has been privy to over the last seven years.

This method of knowledge creation is based on the ethnographic approach of lifecourse of the researcher which brings together personal and professional experiences in helping us to understand the connection between the researcher and the researched (Fois, 2017). The lifecourse paradigm allows us to investigate social pathways, turning points, transitions and trajectories (Elder et al., 2003) and assists in shedding light on the reality of crew life through the author's exploration of crew's emotions, memories and locales which shape their life at sea.

Looking at this approach from a business research perspective, the Critical Research paradigm has been utilized to support the investigation of life at sea through the analysis of how existing professional/business practices influence people and communities (Myers & Klein, 2011). This chapter not only illustrates the four identified areas (exploration, captivity, expectation and reality) but also highlights the substantial pressure that crew face in performing their work role, which can trigger emotional and psychological stressors that the nature and culture of the industry brings about, thus allowing us to investigate how such impact can influence development of new business practices on board to address the challenges that crew face in their daily practice.

Luxury Yachting as a Working Environment

A Crew Member's Perspective

It is common to find a multitude of crew members on the same vessel from different nationalities, ethnic and socio-economic backgrounds. A variety of reasons underpin people's decisions to pursue a career in the industry. Some crew come from less affluent countries and families, and work tirelessly to send money home to support their loved ones. It is also a common notion that crew from more affluent backgrounds pursue a few frivolous, almost gap-years in the industry in search of adventure and a chance to save some money before returning to a land-based lifestyle. These two motivational factors are quite distinct and can be aligned with different stages of the hierarchy of human needs (Maslow, 1943). The motivation of the first group lies within the basic physiological needs, while the latter group could be argued to fit the fulfilment of psychological needs within the Hierarchy. These different motivations for working on board yachts bring with it a very different view of perceived crew dynamics and personal acceptable working conditions.

We must note a distinct difference between personal accepted working conditions and Maritime Labour Convention (MLC) accepted working conditions. The latter will be discussed in more detail later in this chapter. The personal acceptable working conditions referred to are more related to individual crew and personal needs and standards they require. Examples of these would be access to privacy, personal time, personal space, time off to spend at home with family, stress levels created from crew dynamics, culture, demands and expectations of owners, financial compensation and hours of rest. From various conversations with crew from different backgrounds, it can be claimed that in the superyacht industry, accommodation and food offered on board a yacht is in most cases of much higher standard and quality than in other maritime sectors.

Personally, the yachting industry could be seen as a place where one gets paid not only for the work that one does, but also for the amount of time and life that is spent on the boat while not being able to see family and even to call them because of the intense work schedule and fatigue. Therefore, personal life presents a major challenge while working on board.

As with other offshore working environments while opening opportunities for exploring the vastness of the world, the yachting industry is an isolated environment. One can argue, that especially for yacht crew it presents a dichotomy of exploration and freedom in captivity. As an example, on board a vessel that went to Cuba most crew never had the opportunity to set foot on land, even though the crew had visas stamped in our passports, most of the crew could only admire the country from afar. There are many examples and stories from yacht crew, theoretically having seen the world and yet practically only experiencing it from afar while constantly spending time in the same limited space on the boat. It is normal in the industry for the crew to travel to numerous exciting and interesting places such as Russia, Finland and the Arctic but a lot of the time, the crew spend the limited time they might get to explore these places, with the same people they live and work with. Yacht crew get to see the world, but there is always a sense of captivity present in the exploration we can do, whether this is overtly expressed or just intrinsic in the social nature and atmosphere.

The acknowledgment of a certain captivity or constraint in the exploration that yacht crew experience should not be misconstrued as a negative notion. Every person on Earth who goes travelling will be subject to certain constraints, whether it is brought on by a lack of time, money or even adventurous spirit. Here, the acknowledgement is merely linked to the strangeness of exploring the world in a bubble. This odd bubble brings with it intense relationships and friendships as crew live, eat, sleep, work and breathe the same air with the same crew 24/7.

Issues Related to Luxury Yachting: Luxury Yachting and COVID-19

The bubble mentioned above brings with it a form of isolation from the real world that requires consideration. Depending on the location of the yacht, crew can spend months being isolated from the outside world. Regardless of whether it is because they are on anchor or in a country such as Saudi Arabia, some crews have spent months without much, if any, contact with non-crew members. Prior to COVID some yacht owners would impose such restrictions on the crew as part of the standard operating procedure of the vessel. During the COVID pandemic it has, however, become a normal operating procedure for the crew to socialize only in the yacht's bubble. As mentioned above, this type of isolated bubble brings with it interesting and varied crew dynamics and relationships. As the world has been experiencing a form of isolation during the COVID lockdowns, there will be a better understanding of the mindset of a crew member who could be on anchor for several months of the year, with no shore leave.

Since the start of the COVID-19 pandemic, yachting as most other industries, has undergone a massive makeover. The adventure of yachting has been replaced with further isolation in foreign countries away from loved ones. Some yacht owners have permanently moved on board their yachts increasing the workload and pressure on crew, who sometimes do not have weekends or days off when owners are on board. A report published by YPI Crew, a yacht crew recruitment agency, in February 2021 describes the impact COVID-19 had on recruitment in 2020. It stated that among the initial response to COVID-19 experienced by the crew were salary reductions, slashed rotations and loss of employment. The report concluded that YPI is waiting to see if in 2021 a further move towards crew rotation will take place as owners were spending more time on board their yachts and worked on board during COVID (YPI, 2021).

Issues Related to Luxury Yachting: Personal Life and Relationships

As yacht crew, one is always on standby. In other words, very few crew members have the luxury of their own time truly being theirs. It is not uncommon for a crew member to be on holiday a long way away from the boat to be asked to return to work early. During weekends one can be called in at any time; this, admittedly, mostly applies to crew who are not on a rotational contract. The notion of overtime does not exist on boats. Although a vast number of boats are run well and this does not apply to everyone, nevertheless there is always a looming awareness that one might have to return to the boat with zero notice. This knowledge in itself brings a sense of captivity as crew know that in some way, their time is not fully theirs. Many colleagues have missed funerals, weddings and other important family events. We serve at the pleasure of the owner and owners expect their needs and the needs of the yacht to exceed any need that crew might have. Often this is not explicitly expressed—but it is part of meeting expectations by tending to the wants and needs of owners and boats.

In conversation with yacht crew professionals, one will often hear crew mentioning 'our owner really enjoys this' or 'my owner looks after the crew well'. It is something that struck me when I started in the industry, I was shocked that someone would refer to someone else as their owner. Although in reality this is definitely not the case, such phrases still speak volumes about the mindset and subconscious experience crew have onboard yachts. Crew willingly referring to someone as being their owner, demonstrates their subconscious acceptance of being owned. The notion of captivity in freedom becomes prominent as it is represented in the daily vocabulary of yacht crew. Every boat has its own culture as the Captain and owner/s dictate how it is run. However, the universal acceptance of ownership in yachting is extremely common regardless the onboard culture.

Further considering the impact of the yacht bubble, yacht crew have all seen and experienced romantic relationships that formed in this bubble that would otherwise, in a normal situation, not have transpired. It is difficult for most people in the best of circumstances to set emotional and personal boundaries with individuals that they work and socialize with. Given the extraordinary circumstances that crew experience together, from utter work exhaustion to adventurous joy, it is understandable that intimate romantic relationships grow in this context. With that, there is also a real danger of some work and personal relationships becoming toxic due to the limited personal space and time. The emotional strain of living and working with a number of strangers can present a significant challenge to crew members. This requires emotional maturity and recognition of the importance of finding a healthy way to regulate and deal with emotional stressors of living in this type of environment. This raises the question of responsibility for managing emotional strain. Is this the responsibility of leadership on board or the individual crewmember to best equip crew to deal with the extra, unexpected strain of live on board?

Issues Related to Luxury Yachting: Mental Health

This leads us to a discussion of mental health on board. In the last three years there has been a noticeable movement from previous and current yacht crew, organizations such as Nautilus, a union for maritime professionals, and the Professional Yachting Association (PYA) to raise awareness of mental health for yacht crew and other offshore workers. The PYA in association with the International Seafarers' Welfare and Assistance Network (ISWAN) established a Yacht Crew Helpline that specializes in helping yacht crew in various languages free of charge. In 2018 ISWAN conducted a study and then subsequently published a Superyacht Welfare Report which states that:

> Work on superyachts provides exciting opportunities for men and women. Working for rich clients can provide interest and job satisfaction, including visiting some beautiful places and enjoying voyages on stunning boats. On the downside are the long hours when clients or owners are on board, the need to maintain perfect poise and deal with difficult situations at times of tiredness, and the need to keep going regardless of one's physical or mental health .Yacht captains have a wealth of experience and many of them will stay in the industry for decades, treading the difficult line of providing standards close to perfection for discerning clients, and at the same time having responsibility for the crew they manage. For them – and for most seafarers on yachts – the plus side hugely outweighs the negative aspects, and this should be remembered when reading this report. This research is an effort to get behind the façade (which is very solid in all areas of the maritime industry) and find out what kind of services might be provided to superyacht crew from established maritime welfare agencies. (Dudzinski, 2018: 4)

This and other programmes and initiatives such as 'Kelly's Cause' have been advertised and implemented across the industry to encourage crew to seek help and learn skills to deal with the demands of their profession. Most superyachts have management companies and are Maritime Labour Convention (MLC) compliant. '*The Maritime Labour Convention, 2006 (MLC) was adopted on 23 February 2006 at the*

10th maritime session and 94th session of the International Labour Conference (ILC) of the International Labour Organization (ILO)' (SRI, 2021).

Privately registered yachts however, have no legal obligation for compliance but personal experience shows that many private yachts opt for voluntary compliance to this International Body that '*sets the minimum requirements for nearly every aspect of working and living conditions for seafarers, including recruitment and placement practices, conditions of employment, hours of work and rest repatriation, annual leave, payment of wages, accommodation, recreational facilities, food and catering, health protection, occupational safety and health, medical care, onshore welfare services and social protection*' (SRI, 2021).

It is unfortunate that there are currently no published data on the percentage of private superyachts that are voluntarily compliant, but it is common enough practice in the industry that statements such as the following, regularly appear on Yacht Management websites: '*Pleasure Yachts are not subject to the MLC unless they choose to voluntarily comply with the convention*' (MLC, 2013).

This goes a long way to help and ensure crew mental health, as it sets out the minimum requirement for seafarers in terms of various aspects of life on board. One of the most important aspects of ensuring mental health on boats is setting the minimum hours of rest that crew members are legally entitled to. Although this is now regulated by the MLC, in reality when the boss is on board it becomes really difficult to stay compliant. The standards are extremely high and we all know and love being part of keeping them that high. Having said that, it requires time and hours spent working to achieve this. As an example, housekeeping are required to keep guest cabins 5-star hotel check-in ready at all times. From a deck department perspective, the yacht exterior has to be clean from salt marks/water marks at all times, and from the Chef's perspective to serve and create Michelin star quality cuisine coming from a yacht-sized galley with possible limitations in provisioning fresh produce in certain remote places. During a short trip of two weeks it is easy for crew to continue working with minimal to no time to yourself and fewer hours of rest and sleep. During a three-four months' season however, it becomes emotionally and mentally extremely difficult. This is reiterated in the ISWAN (2018) report:

> Several respondents to our survey remarked that they were not sure what form mental support would take and doubted any cost would be covered by insurance. Many assumed that mental health problems must be rife among colleagues, given the conditions of yachting they described: long hours, lack of sleep, little down time or time for exercise, little privacy, confined space and often poor communication and dynamics between and among crew, as well as a 'macho' culture in which mental health is a taboo that would likely result in being fired if flagged. (ISWAN, 2018: 15)

A survey such as *The Welfare of Superyacht Crew* and the knowledge gained, led to opening the previously mentioned Seafarer Help Line. This is a massive stride in the right direction as the industry as a whole is changing and making positive, real improvements to help the crew deal with immense pressure and stress that could lead to mental health issues. Voluntary MLC compliance by some Private Yachts is just one example of this.

Luxury Yachting: a Matter of Perspective

Perception from an Outsider Point of View

Yachties often refer to the yachting industry as 'golden handcuffs' to which crew become used to and grow somewhat addicted to the transient and exciting life on yachts. Anything can happen at any moment, for example, yachts get sold, crew can get fired with zero notice and owner's plans can change. Most crew have some money saved, as they have little normal day-to-day living expenses. The salaries however have been fairly stagnant in the last five years looking at data from YPI who do a yearly comparison (https://ypicrew.com/yacht-crew-salaries-2020). The industry has changed with bigger yachts being run more like cruise ships, working on the yacht, and to a large extent for the owner, is a less intimate environment, as there are various crew coming and going. Owners on a small yacht are more likely to know and have a more personal relationship with the crew working for them.

The industry is also saturated with young South Africans, New Zealanders, English and Australians looking for an adventure and a way to see the world while getting paid. This does not necessarily resonate with other professional and hardworking crew member as their motivation for joining the industry is rather different. While for some this is an exciting adventure, for others it is a way of supporting the family. Tales of friends serving and getting to know the rich and famous while earning thousands of dollars in tips, spread like wildfire and fuels the hopes and dreams of thousands of young, inexperienced yachties commonly known as greenies. The industry is often portrayed as an attractive 'get rich and party industry', which is not the reality for professional crew building a career. The reality is that a lot of greenies never find work. The influx in young and willing crew members however, have had a negative effect on average salaries. As supply increases, salaries decrease as there are enough crew willing to take the job for a chance to gain experience. Additionally, officers from the commercial industry often seek to enter the yachting industry. However, crew who come into yachting from a commercial background do not always understand the standard of service expected on a superyacht and there is a definite learning curve that needs to occur for successful and professional transition.

Tighter tax regulations have also been implemented by several countries which has moved the industry into a much lower earning bracket than before. The expectation of high earnings and no tax paying is slowly shifting, making working on luxury yachts more similar to working onshore. The benefit of saving money remains as crew have their boarding and lodging provided for them by the boat. Yachting is an attractive industry for younger people with no family ties and obligations. As the yachting professionals progress through life and start having families, this way of life brings a whole new set of challenges. This is particularly challenging for those crew who have a partner and children, working at sea on a world travelling yacht with an MLC compliant contract allowing them only the minimum leave days per year. This equates to roughly 38 days per year. Time for time rotational roles are one solution to this problem, as well as yachts who have a home berth where partners and families could potentially stay nearby to be in closer proximity.

Perception from a Staff Point of View

Yachting provides incredible opportunity for crew and will open them up to experiences and enriching development opportunities that most people could only dream of. Having worked on luxury yachts for less than seven years, I have seen over 40 countries, have met interesting people, ate delicacies that I could not imagine existing, drank some of the most exquisite wines the world has to offer while having the opportunity to increase my financial position through saving. Yachting is a strange and wonderous world where one becomes part of a new family of crew members and sometimes even the owner. I know various crew who start on an owner's yacht and then progress to working in their company. Some owners, for example, will invite crew to celebrate big events like New Year's or Christmas with them. Crew will go on adventures and will be exposed to experiences that most people only dream about. However, often it comes at a personal cost. Some people thrive in this environment and others hate it but decide to continue to reach their goal, be it financial or otherwise.

Everyone joins the industry with great expectation of days off exploring the most beautiful places on Earth but the reality is sometimes far from the glamorous ideas and ideals. Personally, I love the industry and cherish everything it has offered. However, it is important to understand the impact working in luxury yachting has on the crew's emotional and even physical well-being. Many of us know that they will not be able to do this forever. For female colleagues who would like to have children this is especially challenging, as this is not a sustainable career for women with families. Another perspective on the matter was researched by a British academic Emma Spence who stated in an article published in The Guardian in 2016 that '*The whole industry is completely gendered… The interior crew are women and the deck crew are male. I've come across two female captains in six years of researching the industry, and I know of two chief stewards who are female. The women retire because owners don't want them in the interior of a boat after a certain age – late 30s and you're off*' (Spence, 2016). I am not sure I agree with that statement in its entirety, but she touches on truth as it is known that older women tend to find it more difficult to secure a job in yachting.

Perception from a Yacht Owner Perspective

Yacht owners for the most part welcome the isolation they can experience within exploration. The privacy that a yacht brings with the freedom to explore is one of the most valuable aspects of owning a yacht. Yacht owners do often enjoy the social life in the Mediterranean but there is also the allure of an environment that is 100% in control of the owner, set up in the most personalized of ways that can travel to the most beautiful and secluded parts of the world.

Referencing Spence's research (2016), yachting involves romanticized notions of escape, it is also rooted in owners' preoccupation with bragging rights. She crewed on one yacht that was sailing around the world; the owners would just fly in when it reached a certain port of call.

> They would leave it in the Med then fly down to Mauritius, then would fly back again when it's in India," she says. "You're really using it [the yacht] as a floating hotel. They would eat meals on board and pop off to do a bit of shopping. (Spence, 2016)

Experience and various industry professionals have taught me that owners experience a certain type of orchestrated reality when they go to these exquisite places. The crew will choose and clear the best paths and prepare ways to get to the best restaurants with the best service. This is obviously to be expected, but to a large extent it also means that it is rare for owners to really experience or immerse themselves in the culture and local lifestyle of a place that they visit. The expectation I have observed and have come to understand is that for most owners it is more an escape from reality and everyday stressors than anything else. Therefore, the expectation is a form of alternate reality of utmost comfort, luxury, beauty and ease. Thus, yachting for them is not necessarily about destinations but about the experience of all their troubles melting away as they step on board.

The best and most professional crew members I have worked with are those who have the ability to anticipate needs rather than just tending to them. It is a crew's job to provide service to the highest possible standard and therefore solve problems even before they arise. It is often one event that is completely out of human control that can make or break a yacht getaway. Throughout the years I have heard several crew members discuss how the weather and the unexpected shift of it, ruined an otherwise perfectly good boss trip. Sometimes crew members will even be blamed for the rain or wind shift as we strive for perfection and are expected to know and plan ahead for the continuation of excellence. Professional yacht crew take great pride and pleasure in their ability to plan ahead and solve potential problems before and as they arise. We are service professionals who are proud of our skills and like to demonstrate and achieve excellence. We thrive when we put our guests' needs above our own and ensure consistent incomparable service in this almost alternate reality that we provide to our guests.

In the end, travel like most things in life is subjective. Everyone travels in a different way with different expectations and goals. Yachting and the type of travel experiences it brings for both yacht crew and owners is seen through a very specific lens, that brings with it both freedom and constraints. There is a certain removal from reality, especially for owners, and that might just be the reason why it is such a popular and ever-expanding industry for those who can afford it. The limitation comes with detachment from real, local experiences.

Luxury Yachting and Tourism

Benefits of Luxury Yachting in Local Tourism

High net worth individuals that visit any place, will bring with them some economic benefit. Yacht owners, for example, will often rent hotel rooms and pay dockage fees to local hotels that will generate income for the local tourism sector. Owners and crew alike are in some instances avid social media influencers and could potentially bring attention to tourist destinations by posting beautifully taken pictures. A large part of coming on board a yacht and exploring a place is also enjoying local cuisine. Owners will often go out for lunch and dinner at local establishments, benefitting the local industry. Chefs also need to provision and source fresh local ingredients, as far as possible to avoid having to fly in fresh food.

Boat repairs often happen right before, sometimes during or immediately after boss trips. This is an economic benefit to the local boating and tourism sector. Yacht crew will often get a little taste of specific places they explore with the owner and might decide to return on their own capacity on leave. All these elements of luxury yachting undoubtedly add to the local economies of destinations during their travels. The tourism sector will therefore benefit economically in various ways, from dockage and port fees to local commerce for provisioning and other stakeholders such as local agents who coordinate with local authorities for yacht, crew and guest movements as well as new health requirements.

COVID-19 has brought an interesting new bubble experience to the yachting sector and could potentially impact the industry in a positive way as contact points for possible infection can be decreased while guests can still enjoy a luxury holiday in an exotic location.

According to a BBC article on the 18 May 2021, 'wealthy people have so far spent more than £1bn on superyachts in 2021 as they seek to escape COVID lockdowns and travel restrictions'. This resonates with Boat International which claimed that the trend towards buying superyachts, which started last summer, was the 'hottest sales streak on record' (Boat International, 2021). This makes us appreciate the growing popularity of this type of tourism which is becoming more popular, although remains inaccessible to most.

The Future of the Yachting Sector as a Tourism Contributor: Concluding Thoughts

This chapter has discussed life at sea through the lens of luxury yachting as a sector of a broader maritime industry. It addressed the four areas which shape daily practices of luxury yacht crew, focusing on exploration, expectations, captivity and reality. Based on personal observations of the author and colleagues, and utilizing the lifecourse framework, this chapter has made the following main observations. Life at sea offers exciting and unique opportunities for luxury yacht crew who are exposed to global travel, professional growth opportunities of serving high-end customers and exciting, albeit challenging cross-cultural teamwork. At the same time, the negative elements include psychological and physical stress of long working hours and severe limitations on personal space and time. Thus, the juxtapositioned reality of global living in a constrained environment presents crew with a complex blend of experiences, emotions, professional tasks and demands which encompass the four areas addressed in this chapter. Exploration, expectation, captivity and reality are tightly intertwined in the world of yachting.

Some recent developments in the industry demonstrate our growing awareness of such challenges, for example a strive to achieve a more balanced life for crew through introduction of Maritime Labour Convention is a step in the right direction. The industry which is developing rapidly and gaining more popularity under the impact of COVID is likely to strive further as owners and guests seek to find an alternative reality away from the pandemic. This demand will generate a wider range of employment opportunities for crew and further positive impact on the economies of yachting destinations.

This chapter contributes to our further awareness of the industry. Linking the experience of the author and her colleagues with their experiences at sea provides a valuable insight into the life which is generally concealed from the general public, thus offering the reader a chance to understand the world of luxury yachting better. From the Critical Research perspective, further studies are needed to identify how business practices on board luxury yachts can be developed to address the issues highlighted in this work, such as mental health of the crew and physical stress. Links to and impact on local economies of destinations could also be researched further to facilitate more effective and innovative ways of collaboration between the luxury yacht industry and destinations in achieving their mutual goals.

References

BBC. (2021). *Superyacht sales surge as wealthy seek Covid escape*. Retrieved June 28, 2021, from https://www.bbc.co.uk/news/business-57158371

Boat International. (2021). https://www.BoatInternational.com

Dudzinski, A. (2018). *The welfare of superyacht crew*. Retrieved June 28, 2021, from https://www.seafarerswelfare.org/assets/documents/resources/The-Welfare-of-Superyacht-Crew.pdf

Elder, G., Jr., Johnson, M., & Crosnoe, R. (2003). The emergence and development of life course theory. In T. Mortimer & M. Shanahan (Eds.), *Handbook of the life course*. Kluwer Academic Publishers. Cited in Fois, F. (2017). Understanding ethnography through a life course framework: A research journey into alternative spiritual spaces. *Area, 49*(4), 421–428. Retrieved June 30, 2021, from Understanding ethnography through a life course framework: a research journey into alternative spiritual spaces—University of Aberdeen (exlibrisgroup.com)

Fois, F. (2017). Understanding ethnography through a life course framework: A research journey into alternative spiritual spaces. *Area, 49*(4), 421–428. Retrieved June 30, 2021, from https://doi.org/10.1111/area.12332

ISWAN. (2018). *The welfare of superyacht crew.* Retrieved June 28, 2021, from https://www.seafarerswelfare.org/assets/documents/resources/The-Welfare-of-Superyacht-Crew.pdf

Maslow, A. H. (1943). A theory of human motivation. *Psychological Review, 50*(4), 370–396.

MLC (2013). *Maritime labour convention.* Retrieved June 30, 2021, from https://www.sarniayachts.com/crew/maritime-labour-convention/

Myers, M. D., & Klein, H. K. (2011). A set of principles for conducting critical research in information systems. *MIS Quarterly, 35*(1), 17–36. Cited in Venable, J. (2011). Incorporating design science research and critical research into an introductory business research methods course. *Electronic Journal of Business Research Methods, 9*(2), 119–129. Retrieved June 30, 2021, from Incorporating Design Science Research and—ProQuest

Spence, E. (2016). *Superyachts and bragging rights: Why the super-rich love their 'floating homes'*. Retrieved from https://www.theguardian.com/lifeandstyle/2016/oct/09/superyachts-and-bragging-rights-why-the-super-rich-love-their-floating-homes

SRI. (2021). *Strategic research for maritime industry*. Retrieved June 28, 2021, from www.Seafarersrights.org

YPI. (2021). *YPI crew annual report: How Covid-19 impacted yacht recruiting in 2020*. Retrieved June 28, 2021, from https://ypicrew.com/2020-yachting-recruitment-data/

Part III

Perspectives on Context

9

Geo-Marine Tourism Development for Luxury Yacht Tourist Markets: Indonesia's Post-COVID Recovery Strategy

Janti Gunawan, Aise Kim, Amien Widodo, and Muhamad Azlian Kautsar

Introduction

The Indonesian government has selected tourism as a strategic sector for the country's economic and socio-cultural development (Indonesia Ministry of Tourism and Creative, 2015; Piccinno & Zanini, 2010). Indonesia's tourism competitiveness has enabled the country to be recognized as one of the popular tourism destinations in Asia, with progressive tourism policies (being ranked 10th), cheaper price for travelling to Indonesia (being ranked 6th) and its natural resources and attractions (being ranked 17th), out of 140 countries (World Economic

J. Gunawan (✉)
Department of Business Management, Institut Teknologi Sepuluh Nopember (ITS), Surabaya, Indonesia
e-mail: Janti_g@mb.its.ac.id

A. Kim
Business School, University of South Australia, Adelaide, SA, Australia
e-mail: Aise.Kim@unisa.edu.au

T. Gladkikh et al. (eds.), *Luxury Yachting*,
https://doi.org/10.1007/978-3-030-86406-4_9

Forum, 2019). Indonesia's unique natural environment comes from being comprised of 17,504 islands with an extensive array of natural resources and diverse geological landforms (Badan Pusat Statistik, 2020). The Ministry of Marine and Fishery records that Indonesia is an archipelago country with 2087 million hectares of marine, coastal and small island conservation areas, with nearly 100,000 km of coastline (Asvaliantina, 2019). This unique natural environment comprising of small islands and coastal areas provides competitive attractiveness to visitors who are motivated to explore new adventures and fulfil their curiosity for learning more about different forms of natural and coastal environments (Tegar & Saut Gurning, 2018).

The marine tourism industry in Indonesia has gained its popularity as one of the special forms of tourism to international tourist markets, attracting 10 million foreign visitors. Of these four million foreign visitors, including luxury yachters, come through seaports and yachts in major marine tourism destinations (e.g. Tanjung Kelayang, Mandalika, Wakatobi or Morotai). The number of international tourists to these destinations has been continuously increasing with stronger growth rates each year, ultimately, leading to a rapid expansion of the yachting and cruise industry (e.g. 5000 yachts—mainly sailing yachts and luxury yachts; and 800 cruises visited the Indonesian ports in 2019) (Asvaliantina, 2019). In order to accommodate further expansion of yacht-based marine tourism, considerable efforts were made by the Indonesian government to improve marine infrastructure which facilitates launching, berthing, staying, servicing, sheltering and keeping the yachts safely and ready to use (Piccinno & Zanini, 2010). In addition, Destination Management Organizations were appointed, and

A. Widodo
Department of Geophysics Engineering, Institut Teknologi
Sepuluh Nopember (ITS), Surabaya, Indonesia
e-mail: amienwidodo@geofisika.its.ac.id

M. A. Kautsar
Marine Engineering Department, Institut Teknologi
Sepuluh Nopember (ITS), Surabaya, Indonesia

budgets were allocated to support the development of additional tourism facilities, tourism marketing, human resource development and better accessibility to destinations (Indonesia Ministry of Tourism and Creative, 2015).

However, the COVID-19 pandemic has threatened government initiatives to achieve the development goals for marine tourism destinations, while restricting foreign yacht visitors. The impacts of COVID-19 crisis in 2020 have severely wounded the local economies of the marine tourism destinations with serious consequences due to regional lockdown that has led to the sudden decline in tourist numbers, loss of local employment and closing down the tourism business sectors. For example, the number of foreign tourists arriving at Indonesia seaports dropped significantly compared to 2019–2020, by almost 77% (see Fig. 9.1).

In 2017, 270 million domestic visitors generated more than 75% of nature-based tourism sector income, and 20% of them are contributed by marine tourism (Indonesia Ministry of Tourism and Creative Industry, 2017). Furthermore, in 2018, there were nearly 12,000 foreign yacht visitors to Indonesia, who spent USD 315 million during average stay of 180 days (Indonesia Ministry of Tourism and Creative Industry, 2020). Although the number of foreign yacht tourists was lower than that of domestic tourists, their contribution to the local economy was about 230 times more than that of the local tourists.

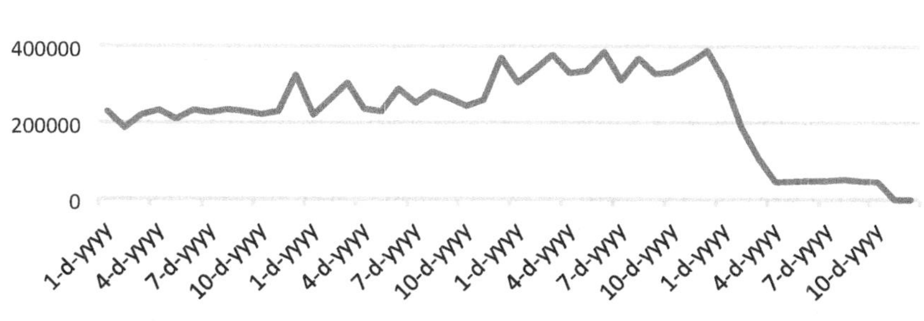

Fig. 9.1 Number of foreign tourist arrivals at Indonesia seaports (person) and the changes (percentage), 2017–2020 (*Source* Badan Pusat Statistik, 2020)

The current socio-economic crisis caused by the impact of COVID-19 forces Indonesia's tourism industry to find alternative options for the development of sustainable marine tourism markets by moving away from traditional marine tourism to expanding the new niche markets. Indeed, marine tourism can be developed incrementally, responding to a new change in tourism demand, while balancing the need of environment-economic and social development for a sustainable growth of the industry (Piccinno & Zanini, 2010).

This chapter aims to present an exploratory study on how geo-marine tourism has been introduced as a new approach to sustaining marine tourism in Indonesia, targeting diversified tourism markets, expanding from luxury yacht visitors to domestic tourists. A SWOT (the strengths-weaknesses and opportunities-threat) analysis has been applied to examine the current development situation of marine tourism destinations. A conceptual model is developed and assessed through an expert judgment method. This method is suitable when relevant data is unavailable (Werner et al., 2016).

This chapter has two major sections. The first section starts with introducing the overview of yacht-oriented marine tourism destinations in Indonesia and its SWOT analysis. The second section focuses on the development of a collaborative geo-marine tourism model as an alternative tool for the post-COVID pandemic recovery situation.

Indonesia Marine Tourism Destinations, Yacht Tourists and the SWOT Analysis

Most of the tourism destinations in Indonesia are located on small islands where the coastal and natural environment is a main pull factor for its growth. Traditionally, Indonesia marine tourism stresses 3S (sun, sand and sea). However, emerging tourist markets seek for satisfaction beyond the 3S, with a specific focus on local attractions across multiple areas including sightseeing to sports and social activities such as scuba diving, snorkelling, windsurfing, fishing, observing marine mammals, kayaking, sailing, motor yachting or joining maritime events, touring fisherman villages, lighthouse, maritime museums and many more (Porter, 2021).

Such emerging tourism activities attract diversified tourist markets who come to the coastal destinations by flights, land transport, cruise ships, ferries or own yachts.

Recently, yacht tourism has grown consistently as one of the mainstream special interest markets. As an excursive, luxurious, entertaining and sporting type of marine tourism activity, yachting has provided more revenues for regional economic development, while creating additional job opportunities for the local community (Piccinno & Zanini, 2010). Compared to the cruise ships, luxury yachts bring a smaller number of people on board, who prefer personalized tours that require more skilled personnel such as the ability of owners to travel around the neighbouring coastal areas (Sariisik et al., 2011).

In terms of infrastructure, maintenance facilities for luxury yachts are required with logistic supplies for the people on board. In Indonesia, yacht tourism is dominated by foreign tourists who come with their own yachts; this could be either sailing yacht or motor yacht, instead of renting yachts at local destinations. A yacht owner typically spends about 3–6 months in Indonesia as the distances between small island destinations are quite long (Indonesia Ministry of Tourism and Creative Industry, 2020). These yachts are mostly owned by international visitors with high household income or pensioners who travel the world to enjoy their retirement period. Furthermore, some of the yacht owners stay on their yachts for longer visits at Indonesian marinas. These visitors may engage in various secondary yacht tourism activities such as tour programmes, fishing, water sports and scuba diving. These secondary activities associated with yacht tourism provide local residents with job opportunities as they offer relevant services to yacht owners such as shipyard, vessel maintenance, additional rental yacht/boat tour services as well as food and beverage services. The capacity and quality of yacht manufacturing and marina-related services in a destination has indirect effects on boosting the successful expansion and efficient management of yacht tourism (Sariisik et al., 2011).

Realizing its huge potential, the Indonesia Ministry of Tourism and Creative Industry commits to promote yachting-oriented marine tourism by conducting regular yearly yacht rallies (Wonderful Sail to Indonesia) and establishing an online process of customs, permits, immigration, quarantine, port service and integrated offline services at several destinations https://yachters.beacukai.go.id/, to streamline the process and attract more yacht tourists. The government also continuously updates the map of diving and surfing locations and entry ports for yachts and cruise liners. The yacht rallies involve both the well-established and new destinations. With the government-led investment projects and tourism assistance programmes, the benefits of Indonesia's yacht tourism are set to expand gradually in the future due to the increasing yacht visits to small islands. In 2019, the targeted demand for Indonesia's yacht tourism featured 5000 yachts, 800 cruises and 100 marinas; while receiving about 4 million foreign visitors to Indonesian coastal areas (Indonesia Ministry of Tourism and Creative Industry, 2020).

The Indonesian government has also implemented a large variety of investment projects in the development of yacht infrastructure and facilities to stimulate multiple effects of yachting-oriented activities at a primary level, the yacht tourism, and at the secondary level, marina facilities and tour package development (Moreno & Otamendi, 2017), attracting both public-led and foreign investors. At the local level, Indonesia's government launched village-owned enterprises regulation in 2015 (Regulation number 04/2015) to promote local entrepreneurship. These enterprises are co-owned enterprises for the local government at the village and the community, formed to achieve three aspects of sustainable development goals: socio-economic outcomes for rural development and environmental management practices for environmental conservation. As a result, yachting-oriented tourism is standing out with its compatibility with the regional marine resources, new development of marine infrastructure and the current government's policies. It also helps create jobs at the rural levels and improve the local economy (Zubaidi et al., 2020).

Although yacht tourism has been recognized as one of the successful tourism sectors in Indonesia's marine tourism destinations, however, there are many challenges and weaknesses in the development of a sustainable form of yacht-oriented marine tourism. In particular, there is unequal development of marine infrastructure, limited skills, lack of local capacity for offering tourism services and ongoing environmental problems (Indonesia Ministry of Tourism and Creative Industry, 2020),

Table 9.1 SWOT analysis of yacht marine tourism development in Indonesia

SWOT analysis	Analysis of yacht marine tourism development
Strengths	• Diverse natural resources and unique coastal environment • Large population in Indonesia, with high youth mobility between islands, creating large domestic markets, and growing middle-class income • Connected large and small islands through various modes of transport • Information technology including the Internet, mobile communication with networks and satellites covers majority of the islands • Custom and immigration platforms applicable to 93 ports in the country, which eases the entry of international yachts into Indonesia (https://yachters.beacukai.go.id/) • The Ministry of Tourism and Creative Industry continuously updates the map of diving and surfing locations and entry ports for yachts and cruise lines • Before COVID-19, the government had held an annual sails event to attract international yachts rally and race, which also involved new destinations
Weaknesses	• Lack of tourism facilities and skills that meet international standards in small islands • The quality of infrastructure varies from one place to another
Opportunities/Potentials	• The rise of new domestic markets, especially youth and middle income • The awareness of the need to improve and the developing digital skills of local tour operators • Potential for product differentiation, with emphasis on unique geographical, environmental and sociocultural values • The coasts are suitable for small yachting activities • Support for village-owned enterprises and their activities with opportunities for jobs and income for communities in remote coastal areas
Threats/Challenges	• Lack of environmental management practices, especially waste management, which threaten the sustainability of the tourist destinations • Managing the impacts of COVID-19 crisis on the local community

which limit the ability to attract luxury yacht visitors to some of small islands. Along with these challenging issues, the unexpected COVID-19 crisis has seriously affected many tourism destinations around the world and in Indonesia; causing socio-economic problems such as increasing poverty, gender inequality and development gaps, which have risen as major threats at a global scale (United Nations, 2020).

As seen in Table 9.1, the SWOT (the strengths-weaknesses and opportunities-threat) analysis helps to examine the current situation of yacht-oriented marine tourism destinations. This situation calls for the needs for strategic development of marine tourism destinations, instead of simply relying on yachting tourism markets only which have been one of major tourism policies led by Indonesia's government.

Model Development Methodology

This study is exploratory in nature. In order to search for a recovery model for sustainable marine tourism, a qualitative approach that combines SWOT analysis and expert judgment was conducted. A qualitative approach can offer insights (Kaushal & Srivastava, 2021) and justify the reasoning behind a certain behaviour when the study is context specific and limited information is available (Werner et al., 2016). Interviews were conducted with a marina manager, yacht rental company owner, cruise training and placement director, tourism polytechnic director, geological expert, local and district government officer in tourism and tour operators at Bali, Banyuwangi and Lombok were conducted. The interviewees represent the stakeholders of strategic tourism destinations in Indonesia.

A Collaborative Model for Sustainable Geo-Marine Tourism Product Development

Sustainable tourism emphasizes that participatory collaboration among stakeholders can overcome the tourism sector's weakness by focusing on the strengths to reach the potential. Following the principles of sustainable development by the United Nations (2020), we propose a conceptual model for sustainable geo-marine tourism as an approach for building resilience in yachting-oriented marine tourism destinations, while recovering from the COVID-19 pandemic (see Fig. 9.2). This model also suggests key major strategies which could help to deal with managing the COVID-induced crisis and mitigating its socio-economic impacts on the livelihoods of local communities; developing new geo-marine tourism products; boosting competitiveness and building resilience in yachting marine tourism; advancing innovation and digitalizing the tourism ecosystem; fostering environmental sustainability and inclusive green growth; and coordination and partnership to transform tourism and achieve the sustainable development goals (SDGs), following the United Nations' agenda (United Nations, 2020).

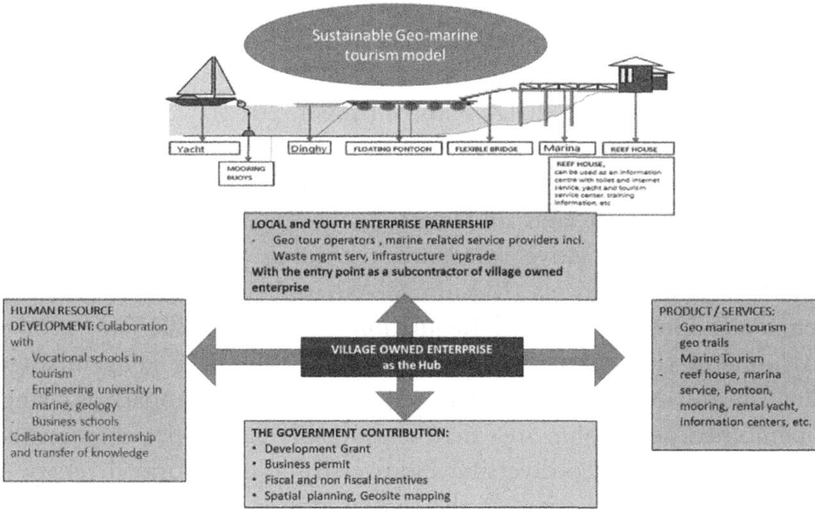

Fig. 9.2 A collaborative approach for sustainable geo-marine tourism (*Source* Developed from the original image of Marine Tourism 2020 [Indonesia Ministry of Tourism and Creative Industry, 2020])

Geo-Marine Tourism as a Form of Product Differentiation for Luxury Yacht Markets

Geo-marine tourism could be differentiated as a new marine product for Indonesia's post-COVID recovery, as it has the great potential for providing yacht tourists with different aspects of tourism experience with a stronger sense of the geological landscapes, which is distinguished from the coastal environment. Indeed, geo-tourism is a relatively new concept in the tourism industry in many destinations. It has developed from a subset of nature-based tourism or ecotourism, which identifies and focuses on the diversity of the geological heritage of the sites (Dowling & Newsome, 2018; Kim & Brown, 2012). Geo-tourism offers distinctive value of the geological environment and an opportunity for new exploration of a geo-site from various perspectives (with the local geology, culture or social lifestyle) which have often been neglected compared to the biological environment in nature or coastal tourism destinations (Dowling & Newsome, 2018; Kim & Brown, 2012).

Geologically, Indonesia is gifted with various unique geological forms as its positioned at the equator with a tropical climate. It offers high biodiversity both on land and water. It is also located at the ring of fire, where it holds active volcanoes, providing a variety of geological attractions and landforms (Masum & Akbar, 2019). In 2019, the president of Indonesia launched Regulation number 09/2019 on geopark development (Peraturan Presiden No. 9/2019 Tentang Pengembangan Taman Bumi, 2019), which stresses the function of a geopark as a single or combined geographic area that has valuable landscapes characterized with geo-heritage, geo-diversity, biodiversity and cultural diversity, to promote tourism activities. At the coastal area, a geopark is known as a geo-marine tourism destination, aiming for protecting marine environment, geo-oriented landscapes, while providing tourists with marine tour activities via ferries, boats or yachts.

Most of Indonesia's geo-marine destinations are in the infancy stage. Yet it has the great potential for providing unique tourism experiences to luxury yacht tourist markets. Niche tourism in a small luxury yacht, that involves small groups of visitors would be more appropriate for Indonesia marine tourism situation, while also being able to contribute to socio-environmental sustainability at a local level. That is, Indonesia's marine destinations may start to serve its luxury local yachters, with smaller size of vessel than international vessel as for the Post-COVID recovery strategy. The small luxury yacht may not fit for ocean passage, but adequate for sheltered coastal waters (Royal Yacht Association, 2020), which are the most features of Indonesian coastal areas. Thus, unique geo-marine-oriented tour products could fit well with the needs of middle-class luxury yacht tourists who are motivated to observe stunning seascapes and learn more about uniqueness of the local geological environment in geoparks, combined with yachting tourism services of modern standards which are available in yachting destinations (Hallott, 2013). It also can stimulate environmental sustainability efforts for the protection of the geological and coastal environments through geotourism's environmental educational activities (Kim & Brown, 2012), and promotes the implementation of health protocols at the local level (Hakim, 2020).

However, unfortunately, most of the Indonesian coastlines face ongoing challenges with a high to severe risk level of various environmental problems such as natural disasters, coastal degradation, coastal pollutions caused by fish farms/tour boats, and changing usage of coastlines due to increasing population and infrastructure development (Wibowo, 2012). Residents and visitors of the marine tourism destinations need to be aware of the environmental risks posed to the region. Indonesia also needs to improve its measures of environmental impact assessment and various environmental sustainability tools for enforcing greater responsibility for protecting the geo-marine environment among the local community and tourists for the future generation. With a large Indonesian population, numerous unexplored islands, limited capacity of human resources, frequent natural disasters, the process of developing a geo-marine tourism model needs to be conducted carefully, considering economic, social, environment and institutional factors.

The following section demonstrates our proposed strategy on how geo-marine tourism can be developed, by focusing on the strengths, overcoming the weaknesses and realizing the potentials, while avoiding the risks.

Village-Owned Enterprise

The main challenge faced by visitors who visit a remote tourism destination in Indonesia is lack of service providers. A village-owned enterprise can become a tourist service company. Their services may include a geo-marine tour, marina services such as pontoon, mooring, yachts rental, yachts maintenance, seafood services and so on.

The mandate of a village-owned enterprise is to promote local economic development; it employs locals and/or supports local businesses, and establishes or facilitates partnership with businesses outside the area. However, in order to execute the mandate, and be a sustainable company, the village-owned enterprise needs professional and skillful human resources. Given inadequate human resources in many small marine tourism destinations, collaboration among local stakeholders can help to foster the development process for successful tourism outcomes in terms of human resources, facilities and infrastructure and tourism promotion.

Human Resource Development

Human resources are critical for business sustainability. One of the weaknesses of Indonesia's marine tourism destinations is related to the lack of local human resources, especially in remote areas which could limit the provision of relevant tourism services or job opportunities for local people. Furthermore, most higher education institutions are located in the urban areas, where tourism education stresses the skills needed for working with hotel/accommodation providers, tour package development, and food and beverage services. Similarly, engineering education focuses on the skills needed to work in engineering companies. There is lack of exposure for implementing the knowledge gained within coastal community-based tourism or becoming entrepreneurs to tap the potential of yacht tourism or geo-tourism.

Village-owned enterprises may invite vocational schools and universities in tourism, engineering, and business for internships and volunteering opportunities. Such a programme can develop students' skillsets, support fresh graduates for entry-level jobs and promote their career prospects or encourage them to be entrepreneurs and deliver community service. For the development of geo-marine tourism products, for example, geological mapping can be done by students from geophysics engineering; marina design and construction can be supported by marine engineering students; tourism information services, e.g. geo-marine tours, yacht history tours, can be run by students from the tourism department; and general business services, digital promotion and administration can be supported by business students.

Besides the fact that most schools and universities are located in the city, such collaboration may also open networks for promoting remote geo-marine tourism activities in combination with existing yachting facilities and services. These education institutions have wider networks with tourism stakeholders than the remote village-owned businesses. Through collaboration with education institutions, local village socio-enterprises may be able to piggyback their network to reach the targeted luxury yacht markets. For example, the Indonesia stock exchange company members have 442 higher education partners for their business research and community development (IDX, 2019). Such networks are useful for introducing new tourism destinations to the well-established businesses. Inviting public listed companies for the development of geo-marine tour and showing how their participation can contribute to SDGs, such as reducing youth unemployment rate, protecting the environment, improving the quality of education, and strengthening the local institutions may also add companies' value in the stock market.

Collaboration for Local Business Development

Yacht-based geo-tourism development strategy could stimulate the tourism business partnership among the local community network groups. Partnerships between local and other private businesses can combine a diverse range of yacht-based geo-tour package and service facilities which can add unique selling points for the stunning geo-locations, improve customer service, and promote innovation. Especially, local start-ups partnerships are first recommended to foster the innovation and strengthen yacht-oriented services (e.g. yacht exhibitions, reef house, information materials for yacht cultural history or luxurious marina). Ultimately, local youth entrepreneurs and start-ups may redesign and develop upmarket geo-marine tourism products and services (e.g. geo-tours, geo-site mapping, yacht culture tours, upmarket seafood tours, and special events at luxurious yacht marinas), which could be more connected to meeting specific demands of international luxury yacht tourists in the future. Yet, it needs to be aware of personalized needs of other niche markets as domestic yacht tourists' needs would differ from the demands of future international luxury yacht tourists.

Developing diversified yacht-based geo-tours and associated local business partnership would be necessary to fulfil specific needs of domestic yacht tourist markets (Piccinno & Zanini, 2010). The World Bank reports that there were about 52 million middle class in Indonesia who enjoy economic security, growing steadily, and they are having disposable income on non-food items, including tourism. They are also more entrepreneurial and can be the driver for local economic development (The World Bank, 2019). For instance, the local luxury yacht visitors may come by other means of transportation to the location, and explore the local area's marine history, geological landforms and cultural attractions by rental yachts. They may create jobs for local land and marine tour operators. They may also stay in the coastal accommodation, as the geo-tour yacht may not have enough cabin to stay. This approach offers an economic recovery for tourism in a pandemic situation.

Furthermore, the government may support the development of a geo-marine tourism entrepreneurial spirit by providing grant for start-ups, fiscal and non-fiscal incentives, a transparent process for obtaining business permits and providing rules and guidance for geo-marine tour package development, such as through spatial planning and geo-site mapping. Such support can help to ensure the balance of environmental protection, job creation and education for the locals. In addition, government may also facilitate the business matching to facilitate the business partnership between well-established accommodation or restaurant service with the local service providers. Thus, the exercise of serving local geo-yacht tourists can be a pathway towards serving international geo-tourism luxury yachters after the post-COVID pandemic recovery situation.

Conclusion

This chapter uses an exploratory technique to find a new model for Indonesia's marine tourism development and recovery following the COVID-19 pandemic. Pre COVID-19, the marine tourism industry had adopted an outward-looking approach by relying on foreign yacht tourism. During the COVID-19 pandemic, the number of foreign tourists dropped significantly, causing an increase in unemployment, poverty and closure of tourism businesses. This chapter proposes geo-marine tourism as an alternative way to continue the development of the luxury yacht tourism market which reinforces multiplier effects on the local industry and boosts the local middle-class income.

Indonesia's numerous unexplored islands have still untapped market potentials for luxury yacht markets. The new development of geo-marine tourism for the local luxury yachters can help to rebuild the economy for recovering from the pandemic, while developing a sustainable path for future international yacht destinations. Village-owned enterprises play a key role in reaching the market potential. A collaborative approach is suggested to expand the current capacity of yacht-oriented destinations. Partnership with education providers in the areas of tourism, engineering and business can help with the human resource capacity issue. Business partnership with local and private businesses may foster the improvement of infrastructure and facilities for yacht-based geo-tourism activities. For example, Indonesia has a number of upmarket hotels, and they may be invited for a joint venture or subcontracting for serving the accommodation services at coastal areas. This approach can accelerate the development of sustainable yacht tourism and ensure the standard requirements by luxury yachters are met. An empirical study to test the model is needed in future research.

References

Asvaliantina, V. (2019). *Strategi Pengembangan Pariwisata Bahari Berstandar Internasional*. Retrieved May 30, 2021, from https://kkp.go.id/an-component/media/upload-gambar-pendukung/DitJaskel/workshop/gelarwisatabahari/StrategiPengembanganPariwisataBahariBerstandarInternasional.pdf

Badan Pusat Statistik. (2020). *Jumlah Wisatawan Asing ke Indonesia berdasarkan Pintu Masuk, 2017–2020*. Retrieved May 30, 2021, from https://www.bps.go.id/indicator/16/1150/1/jumlah-kunjungan-wisatawan-mancanegara-per-bulan-ke-indonesia-menurut-pintu-masuk-2017---sekarang.html

Dowling, R. K., & Newsome, D. (2018). Geotourism destinations—Visitor impacts and site management considerations. *Czech Journal of Tourism, 6*(2), 111–129. https://doi.org/10.1515/cjot-2017-0006

Hakim, L. (2020). COVID-19, tourism, and small islands in Indonesia: Protecting fragile communities in the global coronavirus pandemic. *Journal of Marine and Island Cultures, 9*(1). https://doi.org/10.21463/jmic.2020.09.1.08

Hallott, A. (2013). *The future potential for developing luxury tourism and hospitality in Lapland Title of report* (Issue November). Retrieved May 30, 2021, from https://www.theseus.fi/bitstream/handle/10024/67399/AndrewHallottLuxuryThesisFINAL.pdf?sequence=1

IDX. (2019). *IDX Factbook 2019*. www.IDX.Co.Id. Retrieved May 30, 2021, from https://www.idx.co.id/media/7717/fb2019-halaman.pdf

Indonesia Ministry of Tourism and Creative. (2015). *Rencana Strategis Pengembangan Destinasi Dan Industri Pariwisata Kementerian Pariwisata Tahun 2015–2019*. Retrieved May 30, 2021, from https://www.kemenparekraf.go.id/asset_admin/assets/uploads/media/pdf/media_1564021534_RENCANA_STRATEGISPENGEMBANGAN_DESTINASI_DAN_INDUSTRI_PARIWISATA_2015_-_2019.pdf

Indonesia Ministry of Tourism and Creative Industry. (2017). *Kajian Data Pasar Wisatawan Nusantara, 53*(9), 1689–1699. Retrieved May 30, 2021, from https://www.kemenparekraf.go.id/asset_admin/assets/uploads/media/pdf/media_1553500574_Publikasi_Kajian_Data_Pasar_Wisnus_2017.pdf

Indonesia Ministry of Tourism and Creative Industry. (2020). *Wisata Bahari 2020. Kementerian Pariwisata Dan Ekonomi Kreatif*. Retrieved May 30, 2021, from https://kkp.go.id/an-component/media/upload-gambar-pendukung/DitJaskel/publikasi-materi-2/h20-2/WisataBahari2020.pdf

Kaushal, V., & Srivastava, S. (2021). Hospitality and tourism industry amid COVID-19 pandemic: Perspectives on challenges and learnings from India. *International Journal of Hospitality Management, 92*(September 2020), 102707. https://doi.org/10.1016/j.ijhm.2020.102707

Kim, A. K., & Brown, G. (2012). Understanding tourist perspectives on geotourism experience. *Tourism Review International, 16*(1), 45–57. https://doi.org/10.3727/154427212X13431568321546

Masum, M., & Akbar, A. (2019). The Pacific ring of fire is working as a home country of geothermal resources in the world. In *IOP Conference Series: Earth and Environmental Science*. Retrieved May 30, 2021, from https://doi.org/10.1088/1755-1315/249/1/012020

Moreno, M. J., & Otamendi, F. J. (2017). Fostering nautical tourism in the Balearic Islands. *Sustainability, 9*(12), 1–20. https://doi.org/10.3390/su9122215

Piccinno, L., & Zanini, A. (2010). The development of pleasure boating and yacht harbours in the Mediterranean Sea: The case of the Riviera ligure. *International Journal of Maritime History, 22*(1), 83–110. https://doi.org/10.1177/084387141002200107

Porter, B. A. (2021). Aims & scope. *Tourism in Marine Environments*. https://cognizantcommunication.com/publication/tourism-in-marine-environments/

Peraturan Presiden No. 9/2019 tentang Pengembangan Taman Bumi. (2019). https://peraturan.bpk.go.id/Home/Details/101626/perpres-no-9-tahun-2019

Royal Yacht Association. (2020). *RYA Stability & buoyancy (E-G23)*. Retrieved May 30, 2021, from https://www.rya.org.uk/shop/pages/products.aspx?product=rya-stability-and-buoyancy

Sariisik, M., Turkay, O., & Akova, O. (2011). How to manage yacht tourism in Turkey: A SWOT analysis and related strategies. *Procedia—Social and Behavioral Sciences, 24*, 1014–1025. https://doi.org/10.1016/j.sbspro.2011.09.041

Tegar, D., & Saut Gurning, R. O. (2018). Development of marine and coastal tourism based on blue economy. *International Journal of Marine Engineering Innovation and Research, 2*(2), 128–132. https://doi.org/10.12962/j25481479.v2i2.3650

The World Bank. (2019). *Aspiring Indonesia—Expanding the middle class*. Retrieved May 30, 2021, from https://doi.org/10.1596/33237

United Nations. (2020, August 1). *Policy brief: COVID-19 and transforming tourism*. Retrieved May 30, 2021, from https://unsdg.un.org/resources/policy-brief-covid-19-and-transforming-tourism

Werner, C., Bedford, T., Cooke, R., & Napoles, O. M. (2016). Expert judgement for dependence in probabilistic modelling: A systematic literature review and future research directions. *European Journal of Operational Research, 258*(3), 801–819. https://doi.org/10.1016/j.ejor.2016.10.018

Wibowo, A. (2012). Kerentanan Lingkungan Laut Tiap Provinsi di Indonesia. *Jurnal Ilmu Dan Teknologi Kelautan Tropis, 4*(1), 145–162. https://media.neliti.com/media/publications/100866-ID-none.pdf

World Economic Forum. (2019). *Indonesia travel & tourism competitiveness index report*. Retrieved May 30, 2021, from http://reports.weforum.org/pdf/ttci-2019/WEF_TTCI_2019_Profile_IDN.pdf

Zubaidi, N., Pratama, R. G., & Al-Fatih, S. (2020). Legal perspective on effectiveness of pre-work cards for Indonesian people. *Jurnal Bestuur, 8*(1), 9–18. https://doi.org/10.20961/bestuur.v8i1.42722

10

Emerging Yachting Tourism and Luxury Policy Mobility: Singapore and the Southeast Asian Extension

Emiel L. Eijdenberg, K Thirumaran, and Caroline Wong

Introduction

Destinations craft tourism policy to create key thrust(s) or themes to characterize and promote staged or natural attractions. The policy is then followed by planning and developments to manage tourist flows, attractions and marketing (Baldacchino, 2020). A destination like Singapore adopts and adapts many development policies gleaned through learned experiences from other cities, as reflected in the exhibits at the Urban

E. L. Eijdenberg · K Thirumaran (✉) · C. Wong
James Cook University, Singapore, Singapore
e-mail: k.thirumaran@jcu.edu.au

E. L. Eijdenberg
e-mail: emiel.eijdenberg@jcu.edu.au

C. Wong
e-mail: caroline.wong@jcu.edu.au

T. Gladkikh et al. (eds.), *Luxury Yachting*,
https://doi.org/10.1007/978-3-030-86406-4_10

Redevelopment Authority Gallery of Singapore (Urban Redevelopment Authority, 2020).

Luxury yachting has never been more important in the tourism circuit. This is because the fear of infectious diseases and the realities of COVID-19 have led to an emphasis on the importance of physical distancing, with activities being restructured with small groups clustered apart (Foster & Suwandi, 2020; Kang et al., 2020) and massified tourism on the decline. The restrictions on commercial passenger flights, cruise liners, restaurants and even ticket lines can sometimes be overbearing, especially when family and friends are travelling together (Devi, 2020; Lee et al., 2020; Mouchtouri et al., 2020). With all these travel challenges, luxury yachting as a tourism mode of escapism has never been any more enticing (Zhang, 2020). The pent-up demand from the super-rich and the employment opportunities in the yachting sector have persisted despite COVID-19 (Boyle, 2020). The nature of luxury yachting is especially suited to small groups of travellers known to each other, affording a form of movement with an anchored site that is either exclusive or limited to isolated beaches usually far removed from crowded holiday islands (Young, 2018).

However, there is seemingly a lack of tourism policy on luxury yachting in Singapore, along with scarce policy mobility literature related to area studies (e.g. Asia) and the discipline of tourism. The lack of a tourism policy on luxury yachting in Singapore has been amplified due to the increased consideration given to this emerging tourism sector. Over the last two decades, the steady development from two to four large marinas for luxury yachting is an example of how these nautical leisure facilities have become more popular with the market extension into the tourism sector (Marina at Keppel Bay, 2020; Sentosa Cove, 2020).

Through the literature review and synthesis of a wide range of scholarly journals, consultancy reports, news databases and onsite observations, this chapter highlights the way in which policies in one part of the world, namely the Caribbean and Mediterranean Europe, are assiduously emerging in similar patterns in Southeast Asia. As such, this study examines the significance and potential of yachting tourism in Singapore using the 'policy mobility' lens that considers the adoption and adaptation of a yachting tourism policy based on local circumstances (McCann, 2011), thereby offering a cultural perspective on urban tourism (Giordano & Ong, 2017). The research is based on an extensive scoping of the development and expansion of yachting marinas in Singapore and their nautical activities.

Policymakers and key industry stakeholders in luxury yachting will find special interest in the review of existing practices, and this may facilitate the reconceptualization of tourism policy towards the luxury segment. Scholars studying nautical tourism might also find the review useful in its interrogation of the interconnected geographies and evidence of policy mobility embedded in luxury yachting.

The next section contains relevant reflections on the literature, followed by an analysis and discussion that reviews Singapore's and Southeast Asia's yachting industry developments and policies. This chapter ends with a conclusion addressing future directions for luxury yachting in Singapore and the region.

Literature Review

There has been extensive research carried out in urban planning concerning policy mobility and policy emulations (e.g. Bunnell & Das, 2010; McCann, 2011; McNeill, 2009). However, limited attempts have been made to adopt a policy mobility approach in the analysis of the mobility of tourism strategies and events (Giordano & Ong, 2017). Nevertheless, in the field of tourism, cities have been criticized for adopting quick-fix policies and copying ideas from other creative cities through the implementation of tourism policies targeting the tourists whom they seek to attract (Richards, 2014). Due to intense competition, policymakers are under pressure to develop innovative policies and best practices that can be applied quickly to local contexts. As a result, many new ideas or policy innovations are quickly transferred from one city to another with increased regularity (Peck & Theodore, 2001; Theodore & Peck, 2001) without much thought given to the relationship between mobility, social exclusion and inclusion (Cass et al., 2005; Urry, 2002).

In reality, the interconnectedness of global businesses and travel has resulted in the circulation of policies from one country to another, which now involves a socially constructed process. The term 'policy mobility' has thus become synonymous with 'local globalness' (McCann, 2011: 120), which involves adaptation under local circumstances. In the case of tourism policy transfer, for example, events can be successfully transplanted from one city to another, which supports a cultural perspective on urban tourism and escapes the problem traditionally associated with the 'serial reproduction' process (Giordano & Ong, 2017: 701). These policies are then further adapted and improvised to suit the local context, as illustrated in the study of the various light festivals throughout the world, conducted by Giordano and Ong (2017). The light festival, which started in Lyon in 1999, has now been adapted and successfully transplanted to many countries, including Singapore.

Through such developments, urban tourism has evolved 'from the serial reproduction of mass cultural tourism, offering more flexible and authentic experiences which can be co-created between host and tourist' (Richards, 2014: 119). This is one of the best prospects that applies to the luxury yachting industry. As with place-based policies that take into account a particular understanding of the global context and its array of consultants, exchanges, visits and political and professional networks that cluster around urban initiatives (Clarke, 2010, 2011; Thirumaran & Eijdenberg, 2021; Ward, 2006), policy mobility for yachting tourism in Singapore needs to transcend international boundaries, and a tourism policy that takes into account the local context must be adopted to recreate a negotiated space that is a reflection of contemporary society.

Within the context of policy mobility, Ashcroft's (2017) study of three cities, namely Bombay, Johannesburg and Singapore, offers insights into the various streams of energy constructing the local scene. Singapore has been viewed as a settler nation, a transitional space and a space of flow occurring on two levels: one—being the level of the global economy and the other the level of the place or the people (Ashcroft, 2017). This predicates the context in which we can examine the luxury yachting landscape by taking into consideration the wider global context. Still, we must remain cognizant of how policy mobility for yachting tourism in Singapore needs to transcend international boundaries that take into account the local context. This entails a look at how nautical tourism ports, especially marinas, are seen as complex facilities offering a wide range of services, thereby producing a higher rate of return on equity (Jugović et al., 2011). These include and are not limited to positions of employment in the construction industry, such as the shipbuilders, the berth builders and other industries associated with the supply of raw materials needed to build the marinas, as well as the hospitality and commercial services rendered through food and beverage (F&B) outlets, such as restaurants, bars and cafes.

Sheller (2003) candidly addresses the historical continuities entangled in a developed West being served by the cultural complexities of the exotic East. Said's (1979) and Bhabha's (1994) seminal works return us to a deeper interrogative perspective on how policy mobility and geographical interconnectedness continue to be part of the narrative of the leisure pursuits of the developed world that is itself embedded in the developing regions in the ongoing globalization project (Wood, 2000). The interplay of social inequalities in interactions at touch points, such as the modern marinas, as Lett Jr. (1983) and Sheller (2003) highlight, are the types of imbalances that have been addressed less frequently in the literature related to the more prosperous Asian or postcolonial contexts.

Hence, three clear currents oscillate within the literature relevant to luxury yachting in Southeast Asian waters. First, luxury yachting is a lucrative economy with all its supporting services and lifestyle development. Second, undergirded by power relations, luxury yachting tourism has its own complexities within exclusive communities and serving hosts. Third, there is the desirable agenda of the geo-physical protection of islands and marinas to accommodate the interlocking elements of sun, sea and sand leisure pursuits. This chapter further expounds on the broader global context in which the luxury yachting industry is situated in the historical development of yachting in Singapore. While a pronounced yachting tourism policy is not forthcoming in Singapore and the Southeast Asian region's tourism sector, there is a developing project of experiences mirroring the Mediterranean and (more so) the Caribbean styles of a yachting tourism economy.

A Short History of Yachts and Yachting

Although the history of pleasure boats dates back to Egyptian times, the word 'yacht' stems from the fourteenth century Dutch word *jaght* (in Modern Dutch: *jacht*). In those days, the Dutch used small boats for chasing pirates, and, at the same time, the rich owners had started using theirs for celebrations and other pleasure-seeking purposes. Ever since this time, our understanding of what separates a boat from a yacht is its purpose: pleasure (American Sailing Association, 2020). Since the fourteenth century, an increasing number of aristocrats, monarchs and nobilities of European kingdoms have owned and used yachts. Later, the Industrial Revolution made it possible for the first non-royal but self-made wealthy upper class to spend downtime on yachts. The technological developments of the Industrial Revolution along with the changing needs and desires of people made yachts stronger, better, more advanced and, perhaps most importantly, increasingly luxurious over time (Bender, 2017; Herreshoff, 2007).

In the past 150 years, significant factors have changed the landscape in yachting: the use of lighter, synthetic materials (e.g. fibreglass) and steel have made their introduction next to the predominant wood, and the source of power has changed from mainly wind (i.e. sail yachts) to steam engines and internal combustion (i.e. motor yachts). Yacht races, the so-called *regattas*, have become more accessible for different social classes and are no longer only for yacht owners, and since the end of the Second World War, an increasing number of privately owned luxury mega-yachts have come to market. To name a few, some of these famous luxury mega-yachts include the 'Savannah' owned by the Geneva-based, Swedish-Canadian entrepreneur Lukas Lundin; the 'Predator' owned by the Russian entrepreneur Iskander Kakhramonovich Makhmudov; and the 'Aquarius' owned by the American real estate mogul Steve Wynn. All of these mega-yachts are longer than 72 m, sail under the Cayman Islands flag and are owned by billionaires. The largest luxury mega-yacht is the 180.61-m 'Azzam', built by Lürssen Yachts for a member of the royal family of Abu Dhabi (Boat International, 2020). In the South-east Asian context, one of the most well-known luxury mega-yachts built by KaiserWerft is the 'Cloud 9' (KaiserWerft, 2020), owned by the Singapore-based Australian billionaire Brett Blundy.

Historical Development of Yachting in Singapore

In tracing the history of yachting to the development of the One°15 Marina in Singapore in the twenty-first century, it is evident that there are associated nautical activities related to horizontal development (such as regattas) or vertical establishments (residential enclaves and restaurants) (Jugović et al., 2011). The development of Marina at Keppel Bay is another example of a world-class docking facility that can accommodate up to 180 m of luxury mega-yachts that take people to Singapore's picturesque waterfront spots and idyllic islands nearby. Its location in the privately owned Keppel Island, with a blend of residential, shopping, leisure, entertainment and corporate properties along with dedicated marina amenities, makes the marina a vibrant, integrated waterfront hub in which to live, work and play.

The development of Marina at Keppel Bay is of historical significance. Beginning in the early thirteenth century, many ships sailed from the Straits of Malacca to the South China Sea via the waterway in Keppel Bay. It was William Farquhar, the first Resident of Singapore, who discovered a 'new harbour' west of the settlement, and in 1848, Captain Henry Keppel discovered the sheltered, deep water harbour known as New Harbour, which would eventually be renamed Keppel Harbour in 1900 (Marina at Keppel Bay, 2020). The current developments at Marina at Keppel Bay are closely tied to the harbour in Singapore that led to the prosperity of the island through docking and shipping activities.

While the development of the One°15 Marina Sentosa Cove signals the coming of age of yachting in Singapore, the arrival of yachting predates to 1826 with the formation of the Singapore Yacht Club, which was subsequently renamed the Royal Singapore Yacht Club in 1919 and the Republic of Singapore Yacht Club (RSYC) in 1967. Going back more than 194 years, the club membership included Sir Stamford Raffles, the founder of Singapore, as well as the Prince of Wales, King George VI and Queen Elizabeth II (Republic of Singapore Yacht Club, 2020b).

The first regatta in Singapore was held on 2 January 1834, with four or five boat races appointed for the New Year's Day event (Dunlop, 2010). Sailing regattas had become a regular, local tradition back in the 1940s, drawing large crowds to the Keppel Harbour area near what is now the heart of Singapore's financial district. In 1956, some RSYC members represented Singapore in sailing events in the Olympic Games in Melbourne (Republic of Singapore Yacht Club, 2020a). The RSYC most notably holds the annual RSYC Regatta, a premier local yachting event open to all local clubs and marinas. The RSYC Regatta is the oldest regatta held in Singapore annually in July (Republic of Singapore Yacht Club, 2020a).

Most of the world's luxury mega-yachts are anchored in famous ports such as Dubai, Cannes, Ibiza Town, Marbella, Monaco, Portofino and Saint Tropez. To date, there are about five other marinas in Singapore, including the Marina Country Club, Raffles Marina, SAF Yacht Club Changi, SAF Yacht Club Sembawang and the Changi Sailing Club (Money Smart, 2019). In the case of Singapore, 'Sentosa Cove', a private island-enclave on Sentosa Island located off the south-eastern coast of mainland Singapore, is where the yacht club and marina are based (Sentosa Cove, 2020). The development of Sentosa Cove dates back four decades when the Master Plan for Sentosa Cove was approved by the Urban Redevelopment Authority (URA) in 1993 for a resort residential zone that included a hotel resort, yacht club and marina (Sentosa Cove, 2020). As Sentosa matured into a leading leisure destination, Sentosa Cove would become the residence of choice for those who enjoyed having an island resort as their playground (Sentosa Cove, 2020). Today, more than 60% of its property owners are foreigners from neighbouring countries such as China, Indonesia and Malaysia, as well as European countries and the United States. Property prices differ depending on the type of unit (e.g. condo, bungalow or waterfront villa) but they usually start at a few million USD and can easily go up to US $20 million (Pow, 2011; The New York Times, 2007).

Promotion of Luxury Living and Yachting

The Sentosa Cove enclave, built on partially reclaimed land from the sea, was designed to follow the renowned glamour of the French, Italian and Spanish Rivieras as well as Southern California. The allure is inspired by harbours such as Cannes and Monaco. The enclave is owned by

Sentosa Cove Private Limited (SCPL) (Pow, 2011). This waterfront gated community of approximately 117 hectares has 2500 housing units; a luxury hotel (i.e. the W Singapore—Sentosa Cove, see: W Singapore – Sentosa Cove, 2020); numerous cafes, bars and restaurants; different amenities including shuttle services to the mainland, children's playgrounds, minimarkets, beauty salons, kiosks and stores; and the One°15 Marina Sentosa Cove, the prestigious private marina and yacht club (One°15 Marina Sentosa Cove, 2020).

One°15 Marina Sentosa Cove (referred to as such given its location one degree 15 min north of the equator) is a membership-based club that provides access to facilities such as the boaters' bar, chandlery, dry stack, fuel dock, gym, laundry service, metred power and potable water supply, pool, yachting school and waste pump-out system. The marina offers 270 wet berths and 60 dry berths and can host yachts up to 61 m. It also hosts business meetings, social events and weddings (One°15 Marina Sentosa Cove, 2020). Built to the FIVE Gold Anchors standard, the highest internationally subscribed marina standards set by The Yacht Harbour Association, the marina has been in operation since 2007 and named Best Asian Marina of the Year seven times since its opening (Marina Industries Association, 2020). It is also Southeast Asia's first 'Fish Friendly Clean Marina', is ISO14001 certified for Environmental Management and was recognized as 'Green Maritime Company of the Year' in 2015 (One°15 Marina Sentosa Cove, 2020). The club is touted as the gateway to a wondrous world of unrivalled luxury and an unprecedented lifestyle destination (Yacht Charter, 2020).

Analysis and Discussion

The Policy Mobility Approach and Yachting Tourism in Singapore

Despite the rich history of yachting in Singapore and the associated nautical activities in the development of the marinas, as highlighted in the above sections, there is seemingly a lack of policy planning and direction in yachting tourism. With four world-class marinas, more than a dozen related clubs, at least two yachting shows hosted and increasing numbers of luxury mega-yachts sighted at the marinas, the island state not only has what it takes to be a major yachting linchpin for luxury travellers, but it is also coming of age into luxury yachting commensurate with its status as the Switzerland of the East and Monte Carlo of Asia (The New York Times, 1973; Yapp, 2017).

The concept of policy mobility relates to how policies and approaches are borrowed or how they oscillate from one part of the world to another. Developed and mature regions such as Europe and the Americas have for decades built marinas to facilitate luxury yachting with related activities such as island hopping and beach tourism (Cameron & Gatewood, 2008; Mega, 2016). Singapore's tropical location, just one degree north of the equator, places it at a very amiable position as a luxury yachting hub for two reasons (Varghese, 2020). First, being geographically located in Southeast Asia, Singapore is in a central position in terms of its distance to neighbouring countries with littoral islands dotting the archipelago. This is important because the luxury yachting lifestyle is shaped by people desiring to stop by isolated islands to jump into the shallow waters for a quick swim, walk around the beach and lay in the sand for tanning, then returning to the boat to go back to harbour or move on to another specific destination (Young, 2018). While these types of lifestyles are similar on the Mediterranean coast and the Caribbean archipelago, the same patterns of leisure activities can be found in Southeast Asia.

Meanwhile, the cruise industry took off with the development of the Marina Bay Cruise Centre in 2012 and the Singapore Cruise Centre (SCC) in 1991. The SCC also manages two other ferry terminals, namely the Tanah Merah Ferry Terminal and the Pasir Panjang Ferry Terminal, and in total, the SCC handles a throughput of over seven million cruise and ferry passengers a year, of which about 950,000 are cruise passengers (Singapore Cruise Centre, 2020). These cruise terminals can accommodate more of the upcoming superyachts. The Singapore Tourism Board (STB) supports the development of the cruise industry in Singapore and in Southeast Asia, and serves as the cruise lead coordinator in the Association of Southeast Asian Nations (ASEAN), where it collaborates with regional neighbours to open up more ports and itineraries to boost the region's cruise offerings (Singapore Tourism Board, 2020). The success of the cruise industry in Singapore augurs well for the future of yachting tourism in Singapore.

The choices of a yacht sailing to a variety of tropical, paradise-like islands in Singapore waters are limited. Apart from the southern islands of Lazarus, St John's, Kusu and the Sisters' Islands, there are not many more options for luxury cruising within a radius of two nautical miles. While it is true that some of these islands with reclaimed land beaches are inviting, luxury yachting cruise as an option is not, however, sufficiently marketed in the national tourism collaterals. The STB appears to believe that high net worth individuals (HNWI) usually have their own network to subscribe to personalized services and do not necessarily reach out in the publicly available brochures that highlight the more mundane but no less significant attractions such as the zoos, safaris, science centres and shopping belts of Orchard Road and Marina Bay Sands. Hence, a survey of national websites such as the STB (www.stb.gov.sg) and Visit Singapore (www.visitsingapore.com) yields only dispersed references and links to businesses offering charter services. One such example found in an online editorial by SG Passion Made Possible (2018) reads as follows:

> *If you'd rather be out and about, hop from the bar to the dock and charter a yacht ride from ONE°15 Luxury Yachting. Soak up the sun while you're out on the bay on a vessel that's all yours for the afternoon, or cruise to Singapore's Southern Islands for an escapade from the busy city.*

Chartering is thus advertized as one of the main aspects of the luxury yachting route to experiencing the high life. Domestically, for more than three decades, the expatriate community has contributed significantly to the economic activities on the island (Kolesnikov-Jessop, 2007). For 15 years straight, Singapore has been the top destination for expatriates to anchor their careers and businesses, with approximately 30% from Europe and America, where wages can often be lower and taxes high (Pow, 2017; Sapsted, 2020). The Asia-Pacific is also becoming a brighter market with emerging economies, and the rising number of millionaires and billionaires bodes well for the luxury yachting business in Singapore (Cocks, 2017; Thirumaran & Raghav, 2017).

The development of the marinas by Keppel Bay and Sentosa Cove is very much the consequence of a policy of globalization that facilitated the integration of multinational companies and expatriates into the city's life. The circulation of transnationals and the commercialization of their lifestyle was a mark of global capital and human mobility in the second half of the twentieth century (Koh & Wissink, 2018). Pow (2017) identifies that while developments such as Marina Bay, Sentosa Cove and Keppel Bay were built for the HNWI, 60% or more of these properties are bought by expatriates. Tomlinson (2019) further observes the attraction of Singapore as a stable country where property is seen both as an investment and a holiday lifestyle. This policy determination of creating a lifestyle with spatial distribution and a quiet promotion is a politically savvy approach to valuing the island to a level of a higher service economy, and luxury yachting is part of that larger mosaic of wealth creation and lifestyle production (Pow, 2017).

Luxury yachting is no longer limited to the super-rich. However, tourism bodies need to be cautious about mass tourism in the luxury segment, and they may want to co-create standards and quality checks within the sector while maintaining marketing efforts in their destination brochures with more weight on luxury yachting. Since a 30–40 square metre space can accommodate at least three to six guests for as little as US $1500 with food and drinks at an estimated three to four hour sail time; this is indeed a reasonable range for a small family or group of friends to partake in such luxury experiences.

It must be poignantly stated that luxury yachting is not limited to the rich or expatriate community in Singapore. Local chartering has also become part of the larger make-up of the yachting sector in Singapore (Singh, 2015). Hence, while the state does not have a pronounced policy on luxury tourism or a luxury sector, what has happened is a democratization of access to resources that the locals in the middle class can partake in. This emerging process of inching towards egalitarianism in luxury services may in the long term serve the non-pronounced luxury policy's objective of creating acceptable and reasonable spaces for the HNWIs. Hence, a policy for yachting tourism in Singapore, if crafted, ought to acknowledge the local context in recreating a negotiated space that is a reflection of contemporary society.

Intra-Regional Competition and Cooperation

Caribbean yachting policy and practice provides an exemplary compass for how Southeast Asia can adapt to its own circumstances in growth and intra-regional exchanges. The archipelagic geography and steady pattern of economic growth make the comparison unavoidable. In this subsection, a brief description of the major characteristics of Caribbean yachting tourism is provided, followed by an account of the current Southeast Asian regional realities and their future prospects.

The 1960s and 1970s saw a growth spurt of luxury yachting in the Caribbean. What emerged was a 'playground' in which sexual ludic and open friendships with strangers were common between charter yachters within a culture embraced even by conservative states like the British Virgin islands (Lett Jr., 1983). Many of the island states benefit from the luxury yachting business primarily because they are able to preserve their cays, refrain from the developments of small islands and maintain key inland attractions with idyllic and pristine beaches. These nautical attractions, along with the wet storage and anchorage provisions to replenish and repair services, have provided the region's yachting industry with a key business advantage (Philips, 2014). Furthermore, many of the island countries have a policy of allowing gated communities by the marinas that provide easy access to yachters from North America or Europe via localized immigration processes, as is the case in, for example, Barbados's Port St. Charles and the Dominican Republic's Casa de Campo (Sheller, 2003; Whitehouse, 2017).

The Caribbean yachting industry is starkly different in its concentration and composition. Many of the yachts at various island countries are rented out to tourists, particularly in the peak summer season. In fact, the Caribbean is an archetype for policy mobility that Singapore can follow. Many of the islands, such as the ABC islands (i.e. Aruba, Bonaire and Curaçao), the Bahamas, Bermuda, the British Virgin Islands and the Cayman Islands, among many others, actually have a powerful interlocution among them, such that luxury yachting takes on the meaning of tourism as the boats sail from harbour to harbour and island stop to island stop. These easy stops between different countries in the Caribbean mean that luxury yachting can be an attractive holiday option where distance and national boundaries are not impediments for stop-overs across the Caribbean Sea crest.

The consultant firm Frank Knight (2019), with its annual report on the super-rich and their lifestyles, has revealed that the number of ultra HNWIs residing in Asia, with a net worth of more than US $30 million, has reached 48,245 individuals. The number of millionaires in the world in 2000 was a mere 13.9 million, and by 2019, the number had dramatically increased to 46.8 million individuals (Credit Suisse, 2019). Table 10.1 shows the number of millionaires and billionaires in Asia and selected Southeast Asian countries. By 2023, with the passing of COVID-19 and continued economic growth, the world can expect Asia to reach more than 6.9 million people in the millionaire category.

Myanmar's Mergui Archipelago (e.g. Tawai Island), Thailand's southern islands (e.g. Phuket), Indonesia (stretching from the Riau Islands in the west to West Papua), the Philippines (interspersed from Pahlawan and Boracay to Cebu in the south), Vietnam (e.g. Danang) and Malaysia (the popular duty-free island of Langkawai, where Malaysia's largest marina is located) contain much uniqueness, with some of these locations offering cultural explorations and others wildlife sanctuary attractions in addition to the usual sun, sea and sand pleasure.

As the choices of paradise-like islands surrounding Singapore are quite limited, there is potential for intra-Asian mobility, especially with the consolidation of the ASEAN Economic Community (AEC) that came into effect in December 2015 with 10 member countries that include Brunei Darussalam, Cambodia, Indonesia, Laos, Malaysia,

Table 10.1 Southeast Asian destinations and number of wealthy residents

Destination	Number of individual millionaires[d]: 2018	HNWI[a]: 2019	Number of individual billionaires[b]: 2018
Asia	5,600,000	48,245	758[c]
Singapore	171,559	3598	27
Indonesia	43,118	756	15
Malaysia	29,272	636	12
Thailand	64,131	631	20
Vietnam	12, 327	142	4

[a]Frank Knight (2019); [b]Forbes (2020); [c]Gilchrist (2020); [d]MGM Research (2019)

Myanmar, Philippines, Singapore, Thailand and Vietnam (Association of Southeast Asian Nations, 2016). Among other important areas of consideration, the AEC 2025 vision for tourism is for ASEAN to 'be a quality tourism destination offering a unique, diverse ASEAN experience, and [will be] committed to responsible, sustainable and inclusive and balanced tourism development, so as to contribute significantly to the socio-economic well-being of ASEAN people' (Association of Southeast Asian Nations, 2016: Executive Summary). Another consideration for governments in Southeast Asia is to share a common approach to permits, common vessel identifiable tags and immigration application process for these luxury yachts to easily sail through each other's sovereign waters. Although ASEAN has not achieved its goals of 2015 to allow people to move within a 'borderless' community, starting with a common nautical tourism policy might be a salient move (Thirumaran & Arumynathan, 2016). Such regional extensions of luxury yachting might create another dynamic economic sector for tourism given the rising numbers of millionaires and billionaires in Asia and the growth of luxury mega-yachts.

The Pacific Asia Travel Association (Pacific Asia Travel Association, 2020) has estimated that in Asia, intra-regional tourism accounted for three-quarters to 80% of tourists' movements. For that reason, it seems relevant for Singapore to cooperate with its ASEAN neighbouring countries in the tourism sector, particularly in nautical tourism and luxury yachting, so that yachters beginning their sailing from Singapore or anywhere in the region can navigate the intra-ASEAN sea boundaries without the hassle of immigration rules and gain access to more beaches to explore.

The charter business in luxury yachting is expected to increase in value from US $15 billion to over US $27 billion in the next seven years (Allied Market Research, 2020). Wealth-X (2019) has outlined the future of the luxury yacht business in terms of the onboard technologies, cabin designs and other demands that can weather any sea and long journeys with a stronger hull. One key consideration that was highlighted beyond these constructs was the desire for yachters for nautical adventures. Not limited to or dependent on the exquisite internal cabin facilities and décor, yachters are also looking for islands with wilderness and discovery, and not just another residential resort or a beach stop-over with inland attractions. The likes of the Antarctic wilderness, the Australian reefs and the Southeast Asian hideaway islands can be more competitive and attractive.

With the emerging paradigm shift from mass tourism to special interest emphasis, such as luxury tourism, the yachting industry will be a definite beneficiary. The legacy of COVID-19 will likely mean a greater emphasis on private group tours based on exclusive interest groups or friends and relatives with physical distance maintained, which otherwise is not possible in a mass tourist type of attraction. Whether attractions or travellers adapt, the luxury yachting sector will have to market itself not just to the super-rich but also to the middle and upper classes.

Conclusion

These critical assessments of luxury yachting tourism in Singapore and the region are a response to observed gap of a lack of a tourism policy on luxury yachting in Singapore, along with scarce policy mobility literature related to area studies (e.g. Asia) and the discipline of tourism. As economic prosperity in the Southeast Asian region leads to the production of more millionaires, the integration of higher proportions of expatriate communities and the adoption of a more progressive policy in promoting luxury yachting as a tourism product, we could see more pronounced and beneficial policy mobility similar to Europe and the Caribbean to emerge in Singapore and Southeast Asia.

References

Allied Market Research. (2020). *Yacht charter market business strategies, technological innovation, trends & top players by 2030*. Retrieved November 20, 2020, from https://www.mccourier.com/yacht-charter-market-business-strategies-technological-innovation-trends-top-players-by-2030/

American Sailing Association. (2020). *History of the yacht*. Retrieved November 16, 2020, from https://asa.com/news/2017/12/16/history-of-the-yacht/#:~:

text=The%20yacht%20is%20an%20invention,celebrate%20their%20returning%20merchant%20ships

Ashcroft, B. (2017). Transnation and the postcolonial city. *Australian Humanities Review, 62*, 46–64.

Association of Southeast Asian Nations. (2016). *ASEAN economic community blueprint*. Retrieved November 14, 2020, from https://asean.org/?static_post=asean-economic-community-blueprint-2025

Baldacchino, G. (2020). How far can one go? How distance matters in island development. *Island Studies Journal, 15*(1), 25–42.

Bender, M. (2017). *A new history of yachting*. Boydell Press.

Bhabha, H. K. (1994). *The location of culture*. Routledge.

Boat International. (2020). *Top 200 largest yachts*. Retrieved November 20, 2020, from https://www.boatinternational.com/yachts/the-register/top-200-largest-yachts--25027

Boyle, S. (2020). *Tears, tantrums and tragedies Below Deck: Sian Boyle exposes the seedy reality of superyacht crews being overworked, abused and facing mortal danger while chasing tips from spoilt guests featured in new TV series*. Retrieved November 14, 2020, from https://www.dailymail.co.uk/news/article-8675433/Tears-tantrums-tragedies-deck-SIAN-BOYLE-exposes-seedy-reality-superyachts.html

Bunnell, T., & Das, D. (2010). Urban pulse—A geography of serial seduction: Urban policy transfer from Kuala Lumpur to Hyderabad. *Urban Geography, 31*(3), 277–284.

Cameron, C. M., & Gatewood, J. B. (2008). Beyond sun, sand and sea: The emergent tourism programme in the Turks and Caicos Islands. *Journal of Heritage Tourism, 3*(1), 55–73.

Cass, N., Shove, E., & Urry, J. (2005). Social exclusion, mobility and access. *The Sociological Review, 53*(3), 539–555.

Clarke, N. (2010). Town twinning in cold-war Britain: (Dis)continuities in twentieth-century municipal internationalism. *Contemporary British History, 24*(2), 173–191.

Clarke, N. (2011). Globalising care? Town twinning in Britain since 1945. *Geoforum, 42*(1), 115–125.

Cocks, T. (2017). *Uncharted waters*. Retrieved November 16, 2020, from https://www.pressreader.com/australia/business-traveller-asia-pacific/20170301/281535110756418

Credit Suisse. (2019). *The global wealth report 2019*. Retrieved November 20, 2020, from https://www.credit-suisse.com/about-us/en/reports-research/global-wealth-report.html

Devi, S. (2020). Travel restrictions hampering COVID-19 response. *The Lancet, 395*(10233), 1331–1332.

Dunlop, P. K. G. (2010). *Three burgees: A history of the Republic of Singapore Yacht Club*. Republic of Singapore Yacht Club.

Forbes. (2020). *World's billionaire list*. Retrieved November 18, 2020, from https://www.forbes.com/billionaires/

Foster, J. B., & Suwandi, I. (2020). COVID-19 and catastrophe capitalism commodity chains and ecological-epidemiological-economic crises. *Monthly Review, 72*(2), 1–20.

Frank Knight. (2019). *The wealth report*. Retrieved November 16, 2020, from https://content.knightfrank.com/resources/knightfrank.com/wealthreport/2019/the-wealth-report-2019.pdf

Gilchrist, K. (2020). *This region minted the most billionaires last year*. Retrieved November 14, 2020, from https://www.cnbc.com/2020/06/30/wealth-x-asia-china-india-singapore-create-most-new-billionaires.html#:~:text=The%20region's%20wealthy%20recovered%20from,by%2011%25%20to%20%242.4%20trillion.

Giordano, E., & Ong, C.-E. (2017). Light festivals, policy mobilities and urban tourism. *Tourism Geographies, 19*(5), 699–716.

Herreshoff, L. F. (2007). *The golden age of yachting*. Sheridan House.

Jugović, A., Kovačić, M., & Hadžić, A. (2011). Sustainable development model for nautical tourism ports. *Tourism and Hospitality Management, 17*(2), 175–186.

KaiserWerft. (2020). *About KaiserWerft*. Retrieved November 12, 2020, from http://kaiserwerft.com/

Kang, M., Choi, Y., Kim, J., Lee, K. O., Lee, S., Park, I. K., Park, J., & Seo, I. (2020). COVID-19 impact on city and region: What's next after lockdown? *International Journal of Urban Sciences, 24*(3), 297–315.

Koh, S. Y., & Wissink, B. (2018). Enabling, structuring and creating elite transnational lifestyles: Intermediaries of the super-rich and the elite mobilities industry. *Journal of Ethnic and Migration Studies, 44*(4), 592–609.

Kolesnikov-Jessop, S. (2007). *Gated gardens: Singapore has basically two kinds of real estate-luxury, or state-run*. Retrieved November 20, 2020, from https://www.visitsingapore.com/editorials/indulge-yourself-in-singapore-a-special-guide/

Lee, K., Worsnop, C. Z., Grépin, K. A., & Kamradt-Scott, A. (2020). Global coordination on cross-border travel and trade measures crucial to COVID-19 response. *The Lancet, 395*(10237), 1593–1595.

Lett, J. W., Jr. (1983). Ludic and liminoid aspects of charter yacht tourism in the Caribbean. *Annals of Tourism Research, 10*(1), 35–56.

Marina at Keppel Bay. (2020). *Home: Marina at Keppel Bay*. Retrieved November 12, 2020, from https://marinakeppelbay.com/

Marina Industries Association. (2020). *Introduction to the Global Gold Anchor Scheme*. Retrieved November 5, 2020, from https://www.marinas.net.au/goldanchor

McCann, E. (2011). Urban policy mobilities and global circuits of knowledge: Toward a research agenda. *Annals of the Association of American Geographers, 101*(1), 107–130.

McNeill, D. (2009). *The global architect: Firms, fame and urban form*. Routledge.

Mega, V. P. (2016). Cities and the creative blue green economy. In V. P. Mega (Ed.), *Conscious coastal cities* (pp. 147–174). Springer.

MGM Research. (2019). *Global millionaire population analysis 2019*. Retrieved November 18, 2020, from https://mgmresearch.com/global-millionaire-population-analysis-2019/

Mouchtouri, V. A., Dirksen-Fischer, M., & Hadjichristodoulou, C. (2020). Health measures to travellers and cruise ships in response to COVID-19. *Journal of Travel Medicine, 27*(3), taaa043.

Money Smart. (2019). *How much does it cost to own a boat or a yacht in Singapore?* Retrieved November 12, 2020, from https://blog.moneysmart.sg/high-life/own-boat-yacht-singapore/

One°15 Marina Sentosa Cove. (2020). *One°15 Marina Sentosa Cove: To be Asia's finest marina and lesiure hub*. Retrieved November 12, 2020, from https://one15marina.com/marina/overview/

Pacific Asia Travel Association. (2020). *Pacific Asia Travel Association*. Retrieved November 12, 2020, from https://www.pata.org/

Peck, J., & Theodore, N. (2001). Exporting workfare/importing welfare-to-work: Exploring the politics of Third Way policy transfer. *Political Geography, 20*(4), 427–460.

Philips, W. (2014). *Towards diversification of the tourism sector: A recreational demand study of yachting and marina services in the Caribbean—United Nations, Economic Commission for Latin America and Caribbean*. Retrieved November 20, 2020, from https://repositorio.cepal.org/bitstream/handle/11362/36618/S20131129_en.pdf?sequence=1&isAllowed=y

Pow, C.-P. (2011). Living it up: Super-rich enclave and transnational elite urbanism in Singapore. *Geoforum, 42*(3), 382–393.

Pow, C.-P. (2017). Courting the "rich and restless": Globalisation of real estate and the new spatial fixities of the super-rich in Singapore. *International Journal of Housing Policy, 17*(1), 56–74.
Republic of Singapore Yacht Club. (2020a). *History of RSYC Regatta*. Retrieved November 12, 2020, from http://www.rsycregatta.org/index.php/about/history-of-rsyc-regatta
Republic of Singapore Yacht Club. (2020b). *Our distinguished heritage going back Singapore's early days as a British colony*. Retrieved November 12, 2020, from https://rsyc.org.sg/history/
Richards, G. (2014). Creativity and tourism in the city. *Current Issues in Tourism, 17*(2), 119–144.
Said, E. W. (1979). *Orientalism*. Routledge & Kegan Paul.
Sapsted, D. (2020). *Singapore still top for expat "liveability"*. Retrieved November 15, 2020, from https://www.relocatemagazine.com/news/international-assignments-singapore-still-top-for-expat-liveability-eca-0220-dsapsted
Sentosa Cove. (2020). *About us*. Retrieved November 12, 2020, from https://www.sentosacove.com/about-us
SG Passion Made Possible. (2018). *Indulge yourself in Singapore: A special guide*. Retrieved November 19, 2020, from https://www.visitsingapore.com/editorials/indulge-yourself-in-singapore-a-special-guide/
Sheller, M. (2003). *Consuming the Caribbean: From Arawaks to zombies*. Routledge.
Singapore Cruise Centre. (2020). *About us: Milestones*. Retrieved November 20, 2020, from https://www.singaporecruise.com.sg/index.php/business/about-us/milestones/
Singapore Tourism Board. (2020). *Home: Industries, cruise*. Retrieved November 21, 2020, from https://www.stb.gov.sg/content/stb/en/industries/cruise.html
Singh, B. (2015). *Ahoy, yacht charters getting more popular*. Retrieved November 19, 2020, from https://www.straitstimes.com/lifestyle/ahoy-yacht-charters-getting-more-popular
The New York Times. (1973). *Singapore, "Switzerland of the East"*. Retrieved November 12, 2020, from https://www.nytimes.com/1973/01/21/archives/singapore-switzerland-of-the-east.html#:~:text=The%20Government%20made%20significant%20progress,1969%2C%20expanded%20rapidly%20last%20year

The New York Times. (2007). *Singapore makes a pitch to draw the wealthy*. Retrieved November 20, 2020, from https://www.nytimes.com/2007/04/26/business/worldbusiness/26singapore.html?pagewanted=all

Theodore, N., & Peck, J. (2001). Searching for best practice in welfare-to-work: The means, the method and the message. *Policy & Politics, 29*(1), 81–94.

Thirumaran, K., & Arumynathan, P. (2016). Destination ASEAN, beyond 2015. In P. Manda & J. Vong (Eds.), *Development of tourism and the hospitality industry in Southeast Asia* (pp. 1–11). Springer.

Thirumaran, K., & Eijdenberg, E. L. (2021). Conceptualising destination membership cards for elite travellers. *Zeitschrift für Tourismuswissenschaft, Ahead of Print*: https://doi.org/10.1515/tw-2021-0001

Thirumaran, K., & Raghav, M. (2017). Luxury tourism, emerging destinations: Research review and trajectories. *Asian Journal of Tourism Research, 2*(2), 137–159.

Tomlinson, P. (2019). *Why crazy rich Asians are snapping up luxury homes in Singapore—Just like billionaire James Dyson*. Retrieved November 19, 2020, from https://www.scmp.com/magazines/style/news-trends/article/3021917/why-crazy-rich-asians-are-snapping-luxury-homes

Urban Redevelopment Authority. (2020). *Singapore city gallery*. Retrieved November 12, 2020, from https://www.ura.gov.sg/Corporate/Singapore-City-Gallery

Urry, J. (2002). Mobility and proximity. *Sociology, 36*(2), 255–274.

Varghese, S. (2020). *Yacht chartering is heading towards an indefinite uncertainty*. Retrieved November 11, 2020, from https://www.traveldailynews.com/post/yacht-chartering-is-heading-towards-an-indefinite-uncertainty

W Singapore - Sentosa Cove. (2020). *Luxury resort living experience at W Singapore - Sentosa Cove*. Retrieved November 12, 2020, from https://www.marriott.com/hotels/travel/sinwh-w-singapore-sentosa-cove/

Ward, S. V. (2006). "Cities are fun!" Inventing and spreading the Baltimore model of cultural urbanism. In J.-F. Monclus & M. Guardi (Eds.), *Culture, urbanism and planning* (pp. 271–285). Ashgate Aldershot.

Wealth-X. (2019). *Luxury yacht market: Five key trends influencing the global yachting industry*. Retrieved November 12, 2020, from https://www.wealthx.com/intelligence-centre/exclusive-content/2019/luxury-yacht-market-trends/

Whitehouse, K. (2017). *Witness says Menendez had access to Dominican paradise at resort*. Retrieved November 10, 2020, from https://nypost.com/2017/09/18/witness-says-menendez-lived-large-at-dominican-resort/

Wood, R. E. (2000). Caribbean cruise tourism: Globalization at sea. *Annals of Tourism Research, 27*(2), 345–370.

Yacht Charter. (2020). *ONE°15 Marina Club: Asia's finest marina at Sentosa Cove*. Retrieved November 12, 2020, from https://www.yachtcharter.sg/one-degree-15-marina-club

Yapp, J. (2017). *Singapore: Asia's Monte Carlo*. Retrieved November 17, 2020, from https://discoversg.com/2017/06/01/singapore-asias-monte-carlo/

Young, S. J. (2018). *The yachting lifestyle: Luxury's latest offerings*. Retrieved November 12, 2020, from https://www.luxurytraveladvisor.com/cruises/yachting-lifestyle

Zhang, J. (2020). *The great escape: Luxury yachting for Asia's richest*. Retrieved November 21, 2020, from https://www.prestigeonline.com/hk/pursuits/motors/the-great-escape-luxury-yachting-for-asias-richest/

11

Re-Imaging Malta: The Potential of Sailing and Yacht Events as a Tourism Niche—a Case Study of the Rolex Middle Sea Race RMSR

Caroline Navarro and Andrew Jones

Introduction

Sailing and yacht racing have seen exponential growth over the last decade and as a sport the Rolex Middle Sea Race (RMSR) has established itself as one of the major annual Mediterranean yachting events which starts and ends on the island of Malta. Identifying its development and growth potential towards assisting Malta's niche tourism market are the key aims of this study.

In this context, this chapter seeks to explore the role of the RMSR as an integral part of the sports event calendar and evaluate how its

C. Navarro (✉)
Independent Luxury Yacht Practitioner, Naxxar, Malta

A. Jones
Institute for Tourism Travel and Culture, University of Malta, Msida, Malta
e-mail: andrew.jones@um.edu.mt

T. Gladkikh et al. (eds.), *Luxury Yachting*,
https://doi.org/10.1007/978-3-030-86406-4_11

attributes can assist in strengthening and contributing to tourism diversification and Malta's niche tourism market growth. The main focus is on addressing the challenges the event faces in relation to its maritime and tourism contexts and exploring the potential benefits and opportunities they accrue. This chapter reflects on the RMSR event since 2012 (a record year with 82 entries) as a specific case. From that time the Rolex sponsorship arrangements has steered the event to a major component of Malta's sailing and yachting sports event calendar (Rolex, 2021).

The research questions focused on evaluating the specific and essential contribution of the event highlighting the issues pertaining to awareness, promotion, branding and socio-economic impacts and to draw broader implications for hosting luxury quality events of this nature. A qualitative methodological approach with purposeful interviewing and sampling methods was utilized in order to inform and underpin the discussion. Results suggested that these tended to include a general lack of statistical data making it impossible to measure the quantitative nature of the event from both an economic and social value, together with the difficulty in the provision of key infrastructure such as berthing facilities which impacted upon prospective business opportunities. In this respect the results also confirmed key benefits and opportunities for developing Malta as an over-wintering destination for yachts and the advantages of investing in sailing and yachting and associated events for diversifying the Maltese tourist economy, thus moving from a situation which is currently dominated by 'mass' traditional tourism markets to one which can find future opportunities for more upmarket, quality-based niche tourist developments.

In this context the RMSR is very much perceived as a luxury event. Its association with the Rolex brand and links to other large sailing events in the Rolex series and the Americas Cup very much support such notions. The concept of 'legitimacy in luxury', which includes an exceptional process, a product of the highest quality and a tradition or history associated with the brand (Vigneron & Johnson, 2004). This very much fits within the constraints of where Malta aspires to place itself regards its brand image.

Developing Niche Markets

Niche tourism has emerged as an alternative to mass tourism and as Robinson and Novelli (2005: 1) suggest 'it implies a more sophisticated set of practices that distinguish and differentiate tourists'. For destinations wishing to change their response towards traditional and newer concepts for sustainable tourism, niche tourism can offer a set of experiences tailored towards the visitor's needs and wants while attracting higher-spending tourists. This is particularly relevant nowadays as niches have been broken down from larger macro-niches like culture, rural and sports to micro-niches for example, sailing and yachting tourism, to suit the needs for differentiating the tourist economy. As pointed out by Macleod (2006) choosing a holiday and destination is partly identity-making. Studies by, for example, Trauer and Ryan (2005) and Trauer (2006) also suggest that individuals 'adjust their needs and desires based on images of societal behaviour that 'promise' societal acceptance.

Therefore, those interested in special interest travel usually choose a holiday in accordance with the interests and activities taking place within home settings. Trauer (2006) also suggests that although event tourism can be the special focus of activity, events can be further specialized by themes and interests such as sport events (for example World Masters, The America's Cup). The emergence of satisfaction, as the principal criterion of vacation selection, has led to a marked shift towards 'active holidays' and towards an interest in 'experience-oriented' holidays for the more sedentary visitor (Weiler & Hall, 1992). These are notions that are at the very heart of new tourism development strategies across the Maltese Islands. They are also especially pertinent as Malta tries to reposition its tourism economy from one which is largely reliant on 'mass' to one that provides a more specialized, sustainable and diverse tourism offer (Visit Malta, 2021).

Maritime and Nautical Tourism: the Blue Economy

The EU's maritime region accounts for about 40% of its GDP and population. Tourism contributes to a substantial part of this impact, employing more than 2.36 million people. As an example, 51% of all hotel beds in Europe are to be found on coasts and maritime environments. Working towards a more sustainable approach to employment opportunities in these regions the EU values the 'blue economy' as an increasingly important means to secure better sustainable maritime tourism economies (EU, 2014). A more recent report (EU, 2016) also identified coastal and maritime tourism as an important subsector of tourism and the largest maritime activity in Europe. Employing over 3.2 million people, the report estimates that this sector generates a total of € 183 billion in gross value added and represents over one-third of the total European maritime economy. In this respect, as far back as twenty years ago authors such as Hall (2001) have identified that 'ocean and coastal tourism are widely regarded as one of the fastest growing areas for contemporary tourism'.

In May 2012, the European Commission representation in Malta also set up a public consultation programme aimed at local Maltese stakeholders to formulate better knowledge on the challenges and opportunities for Maritime and Coastal Tourism in Europe. Respondents classified non-beach recreational tourism such as nautical boating as a very important phenomena. The discussion focused on Maritime Spatial Planning, Integrated Coastal Zone Management and the 'Blue Growth Strategy' (EU, 2012, 2016).

The discussions have been concerned with the environmental sustainability of the European coast, its heritage, issues of seasonality, product diversification and innovation, competitiveness, transport and accessibility, infrastructure and employment opportunities and skills. As part of the initial 2012 study a twelve-week Public Consultation by Member States and stakeholders resulted in the annual European Maritime Day Conference held in Malta in May 2013 (EU Maritime Day 2013). The conference discussed matters relating to coastal and maritime tourism, based on the integrated maritime policy and blue growth strategies.

As such, nautical tourism or that associated with the 'blue economy' provides a focus for new alternatives for Maltese tourism. In this respect the European Union initiatives have presented new opportunities for boosting European coastal and maritime tourism destinations, including Malta (EU, 2012).

The initiatives suggest that nautical tourism/sport can deliver a high-quality experience which can help towards overcoming tourism seasonality and diversify tourism economies. In this respect it can also assist in improving the SME market, create jobs and wealth for coastal communities and increase profitability. In turn, evidence suggests that such tourists will spend on average 30% or more on such activities and provide and help promote a more sustainable approach towards the environment, local community, jobs and customer choices. Again, these are the notions that the Malta Tourism Authority (MTA) wishes to fully capitalize upon, and which in turn ultimately uses 'sailing and yacht-based tourism' as a vehicle to transform Malta as a destination for the sport of sailing, yacht chartering and yacht racing events (Visit Malta, 2021).

The Rolex Middle Sea Race: RMSR Malta

The Rolex Middle Sea Race was conceived in 1967 to celebrate the RMYC (Royal Malta Yacht Club) new premises at Couvre Porte, Fort Manoel on Manoel Island—Malta. It was the result of a sporting challenge between sailing advocates, Jimmy White, Alan Green and the Maltese Ripard brothers. The race aimed to promote a harmonious integration between the British and Maltese which were at the time the two main communities on the island (Borg, 2012a).

The Rolex Middle Sea Race is one of the top three offshore yacht races worldwide along with the Sydney to Hobart yacht race and the Fastnet (Global Yacht Racing, 2012). It is the last classic offshore regatta of the season and as Ted Turner (USA), former CEO of Turner Communications (CNN), once suggested it must be the most beautiful racecourse in the world (Sultana, 2012).

This 606 nautical mile yacht race, similar in length to the Fastnet Race, leaves Malta annually towards the end of October. The race heads north from Malta anti-clockwise along the east coast of Sicily past Mount Etna and through the Messina Straits to the Aeolian Islands and Stromboli and then south down the west coast of Sicily towards the islands of Pantelleria and Lampedusa, with the finish line in Malta (Fig. 11.1).

The first race began in 1968 with only eight yachts participating. Already gaining popularity in the third year, 1970 saw twenty-five yachts participating from seven different countries plus some important names in the sailing scene which helped raise the international status of the race. These early years were important in establishing the value of the race and providing a stimulus for further investment in yachting as a tourist activity and event across Malta (Borg, 2012a). The end of the 1990s saw declining numbers, yet 1999 saw three maxi yachts (yachts above 30 m) participating and renewed interest.

For the following three years the Royal Malta Yacht Club (2012) struggled to fund this top yacht event as it was fast building recognition. In 2000 Rolex sponsored a Rolex Cup event that took place in Malta. This led to the Middle Sea Yacht Race enhancing its status

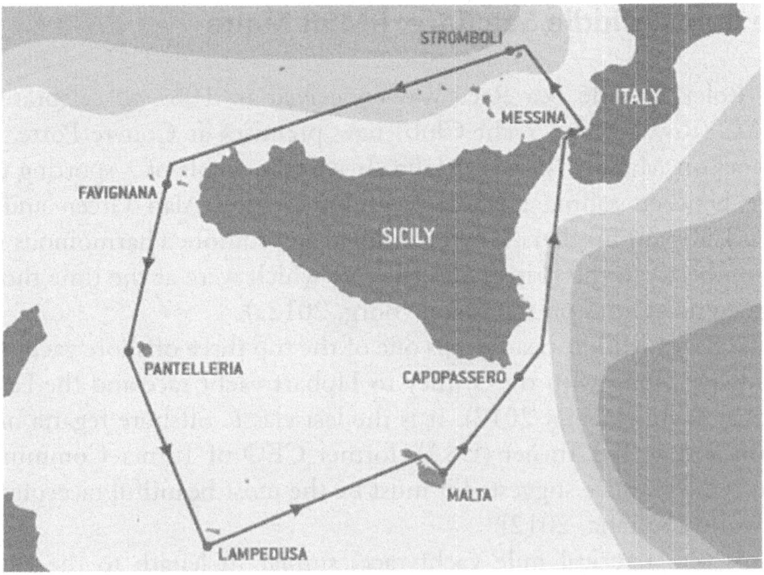

Fig. 11.1 Rolex Middle Sea Race route (*Source* www.rmyc.org, 2013)

through its sponsorship from Rolex in 2002. As a consequence, the exposure of the race now receives some of the world's most renowned yachts and yachting personalities (Borg, 2012a). 2012 saw a record number of entries reaching 82 yachts in total. Five maxi yachts participated with the majority coming from Italy, the United Kingdom and Russia. Line honours was won by the Slovenian Super Maxi Esimit Europa 2 (Rolex Middle Sea Race 2013). Comparing attendance over the past 12 years from 2008 to 2019, participation grew by 45% with number of entrants rising from 78 to 130 in 2018 representing foreign owned and crewed boats (Yachting Malta 2020).

Rolex Middle Sea Race mainly attracts yacht owners and crew. Yacht owners, captains and crew are considered medium to high net worth earners and come with a variety of needs ranging from flights to accommodation, shopping to entertainment. Organizers, media, family and friends of overseas participants often stay for a week or longer since the race begins and ends in the same location. There are often many returnees—crew and yachts every year, as the race prides itself on its unforgettable and hospitable experience. When one returnee participant was asked 'What makes the Rolex Middle Sea Race so special?' he answered 'Friends, a welcoming club, the best boats and sailors around, a splendid Malta, a fabulous start, incredible sights along the course, two volcanoes and challenging seas. At the Rolex Middle Sea Race participating is truly a reward in itself' (Borg, 2012b).

In addition to a sailing view, the race has been termed to be challenging and difficult according to participants who have sailed the course several times (Borg, 2012b). Borg also suggests that due to its changeable and demanding conditions the race has been likened to a combination of races. In this respect, Borg (2012a, 2012b) goes on to suggest that the race offers new and unexpected weather conditions every year and thus is not predictable. He especially points out the unpredictability of the winds and sea currents that change throughout the race. In this context such conditions make the race 'intriguing' and 'challenging' for every skilled yachtsman and thus provide one of the main reasons that entices sailing and yachting professionals to return year after year (Borg, 2012a).

Yachts participate in four classes, ranging from 30 to 100 foot for the most high-tech builds. Throughout history the race has attracted a select number of the most famous yachts, owners and racing crew who have transformed it into a legacy and 'must do' for the hardened yachtsman. In 2012 the race attracted close to 1000 competitors, guests, race officials and media (RMYC, 2012). In 2019 a total of 113 boats took part in the Race, 'a 45% growth over the last 12 years'. 830 non-residents participated in the race, consisting of captains and crew members. 38% had family and friends accompany them, 94% said they would visit Malta again and 47% visited Malta between races. In total the race welcomed 1009 attendees both local and foreign (Yachting Malta 2020).

The RMSR is not only one of Malta's growing tourism events but a very well recognized and sought-after sports event internationally, solely sponsored by Rolex. The RMSR is an invitee-only event, divided up into those taking part in the race and family and friends who wish to support the participants. Each group comes with high expectations of quality in the event management of the RMSR; starting from the crew party at the yacht club two days before boats depart, to the breath-taking start catered for in the historic gardens on the Grand Harbour to the legendary prize-giving in Valletta's grand heritage site, the Mediterranean Conference Centre known as the Holy Infirmary.

Luxury traditionally associated with exclusivity, status and quality. Phau and Prendergast (2001) state that luxury brands exude exclusivity and therefore buying their products is perceived as buying quality. Luxury brand identity is easily recognizable and that helps to sustain brand awareness and clientele while having sustainable sales. The marketing of luxury goods has become increasingly complex, being associated not only with conveying an image of quality, performance and authenticity, but also with attempting to sell an experience by relating it to the lifestyle constructs of consumers is particularly evident in Rolex's choice to market based on the sponsorship of elite sports events attracting the rich and successful particularly through golf tournaments and regatta sailing. This also fits well with recent research by Bladen (2021) which covers what he calls the 'service semiotics' of luxury events which considers primarily the design and delivery of luxury event-attendees' experiences. These are sentiments also expressed by Atwal and Williams (2009) in their assessments of 'Luxury Brand Marketing' who, in turn, imply that 'experience is everything'. The RMSR event certainly corresponds with such notions in both respects.

The growing interest in this form of tourism has also been increasingly accepted by the Maltese government which, from an economic point of view, has recognized superyacht racing as a specific tourism asset of its own. According to Edward Scicluna (an economist and ex-Minister for Finance), average daily spend can be estimated at 5000 Euro per boat, of which the majority is spent on repairs and the remainder on utilities and mooring. Estimates also show that on average crew between 250 and 500 Euros is spent every day on crew. He also suggests that excluding airfares the average daily expense of visitors related to the event is estimated around 350 Euros, which is higher than what the average tourist spends during an entire holiday in Malta. Such sums equate to the possibility that the event contributes more than fourteen million Euros or more to the local tourism economy (Sultana, 2009).

In this respect Scicluna also suggests that The Rolex Middle Sea Race has become more than just a race but an icon among sporting challenges which brings high spending visitors and also extends the summer tourist season. In short, the race comes at the perfect time to attract people to 'over-winter' in Malta which could in turn lead to almost six months of dockage and repairs and associated boat chandlery services, accommodation, hospitality and transport services (Sultana, 2009).

Based on a more recent study commissioned by Grant Thornton for Yachting Malta in 2019 on the Economic Impact of the RMSR on the Maltese economy, 940 people specifically visiting Malta as participants/spectators together with the RMYC spent 2.3 million Euros. An average of 193 Euros per day expenditure for an estimated stay of 11.5 days, 81 Euros more per day than your regular tourist. Owners spent 216 Euros on average for food beverages, accommodation, entertainment and transport for themselves, their crew, family and friends. The PR value of the online articles and TV features rose to 2.6 million Euros bringing the total economic impact to 4.7 million Euros in 2019 (Yachting Malta 2020).

Clearly over the last decade the RMSR has established itself as an important maritime event, the success of which now contributes to a growing maritime tourism sector both within Europe and within Malta.

This very much contributes to current policy for the Maltese Islands (Visit Malta, 2021) and more broadly aligns to European maritime tourism policies which are increasingly linked to the promotion of the 'blue economies' across Europe (EU, 2016).

Research Approach

This chapter is informed by a qualitative study which utilized purposeful interviewing and sampling methods. The research questions focused on evaluating the specific and essential contribution of the event highlighting the issues, opportunities and benefits. The key focus for this study included the following key research objectives.

- To explore and evaluate how yacht events can help towards offering a more quality tourism product.
- To identify and analyse the socio-economic benefits that can accrue from the RMSR event and other potential yacht events for tourism in Malta and the broader maritime economy.
- To ascertain and evaluate the business potential of such events.
- To determine how to capitalize on such events in a more sustainably sound manner.

Primary data collection was based on purposeful sample selection of interviews from race participants and key stakeholder groups associated with the RMSR. This approach aimed to provide distinctive insights and reflections on the event outcomes in terms of both the positives and the negative impacts on the broader tourism and maritime economies. This also provided a unique discourse on lessons that can and may be learnt from the hosting of events such as the RMSR and the wider prospects for sailing and yachting as 'added value' for event and tourism destination management. All 33 interviews were conducted with key stakeholders from key sectors that included race participants, organizers, race representatives, transport representatives, sponsors, community/neighbourhood groups, tourism and hospitality sectors, government departments, local sport/youth clubs, local maritime industries and environmental groups. The variety and diversity of stakeholders selected aimed to provide a unique and diverse representation of broad opinion in order to seek answers to the research questions and research objectives set. The data was analysed through discourse analysis techniques utilizing grid matrices to evaluate and ascertain common themes and responses that occurred from the discussions and questions asked.

Since the RMSR takes place annually in October and forms part of the international yacht racing agenda, attracting yachting enthusiasts from a variety of nationalities with many returned participants has provided an ideal opportunity for gathering data through in-depth interviews during the event. The research topic often best dictates the methodology, and, in this case, an inductive qualitative method was deemed to be the most suitable since the purpose of the study was to explore and provide an in-depth analysis of the research subject based on the research objectives set. This, as already stated, aimed to explore and evaluate the impacts of the RMSR event for Maltese tourism and draw out broader reflections for maritime-based tourism economies more generally.

This study best fits under the illustrative style and particularly in an organizational theoretical structure. It deals with how an event is structured, functions and performs particularly within the different inter-organizational partnerships. Bickman (1987) reflected that theory needs to distinguish between the substance and how it can be implemented effectively.

In short, the research was based on a case study of a real-life event. As such, its academic value could be interpreted as rather specific and industry focused. However, the empirical value and lessons learnt from the enquiry can also be applied to broader tourism contexts albeit with a proviso that the results obtained are very specific to the RMSR. The in-depth individual interviews with the key stakeholders thus provided the primary data which in turn when manipulated using discourse analysis techniques enabled an analytical framework for drawing out key discussions, findings and conclusions.

Findings and Discussion

The initial aim of the research was directed towards exploring the role and impact of the RMSR on Malta's tourism and maritime communities and economies. The findings from the research demonstrated some interesting relationships between the results, stakeholders and the existing literature relating to the topic.

Uniqueness

All respondents agreed that the race offers a very unique experience and that very few races have all the attributes the RMSR can offer. These particularly related to the uniqueness of the Valletta Grand Harbour setting, the value of Malta's tourism offering and transport links, the nature and geography of the event and support infrastructures. The unforgettable start (Valletta Grand Harbour) was mentioned by most respondents, could be capitalized on further from a number of perspectives. This might include, for example, branding and potential media coverage which in turn can attract additional visitors and tourist spend. Such notions and concepts of co-branding have been discussed by Jago et al. (2003), Getz (2008), and Allen et al. (2011).

These authors emphasize the key ingredients for the successful promotion and management of events using case scenarios which go some way to match experiences gained from the RMSR event. In this respect one of the yachting fraternities interviewed stressed the fact that this event should be utilized to brand Malta's image as an alternative and perhaps upmarket tourism destination. Following Hughson (2011), 'sports re-imaging' could apply which would assist in promoting Malta as both an alternative sail and yacht-based tourism destination. Another respondent from the yachting fraternity, also suggested that so much more could be done from a local perspective to capitalize on the event. This, as suggested, might include better promotion, enhanced media coverage and associated mini yacht-related and sailing events hosted by adjacent local municipalities.

Market Recognition, Organization and New Market Opportunities

The uniqueness of the race starting and ending in Malta, thus leaving family and friends on the island for a week was recognized by several respondents as a key advantage. As such this offers Malta the opportunity to capitalize on 'tourism gain' and the development of new business opportunities. Sultana (2009) has already estimated that the average daily spend can be anything between 250 and 500 Euro excluding flights and accommodation. Weed and Bull (2009) have also supported such notions by remarking on the wider tourism benefits from such events. Allen et al. (2011) also suggest that such a major attraction can be used as a foundation for image building as in the case of New Zealand (Kane, 2010). In this respect, New Zealand has established itself as a lead nation in yacht and sailing events. New Zealand sailors have claimed, over the years, most of the top international sailing trophies including the America's Cup and Whitbread Race. At a conservative count, New Zealand yachtsmen have won more than 60 world titles and with 18 medals with yachting now established as the country's most successful Olympic sport. There is no simple answer to the secret of New Zealand's sailing successes but there are some defining factors: dedication from an early age, constant exposure to strong local competition and a highly competent marine industry, which have helped establish such a success story (New Zealand Tourism, 2021). It is experiences, notions and aspirations of such an approach that could be applied very appropriately for Malta's current sailing offer which in turn could provide building blocks for future growth.

Coordination and Communication

Most respondents agreed that in order for an event to benefit from a wider commercial perspective there needs to be better coordination and communication between the government authorities such as the MTA and the RMYC (Weed & Bull, 2009). In this respect, communication between stakeholders appeared to be of prime importance in order to avoid disappointment, embracing opportunity or avoiding weak participation. This has also been an issue and challenge recognized by authors such as Arruda and Ferrand (2007) in their assessments of the ISAF Sailing World Championships in Cascais Portugal. With reference to this, one yachting executive remarked: 'Why does the Fireworks Festival and Isle of MTV - Music Television event get more publicity than the RMSR? (Malta Independent, 2007). Coordination and marketing appeared to be weak areas in the promotion of the RMSR'. In this context, a key yachting executive interviewed suggested that a way forward would be to further analyse and identify the profile of the 'sailing' visitors/participants and develop a branding and marketing strategy appropriate for future market-related development. In support of such an approach the creation of a 'week-long event', perhaps a 'Race Week', sailing schools and regatta events were other suggestions put forward. Authors such as Allen et al. (2011), Weed and Bull (2009) and Getz and Page (2016) have also advocated such approaches and the development of a portfolio of attractions and events that reflect national values, cultures and history. Such an approach could certainly open up new market opportunities for tourism in Malta which could in turn become embedded and help diversify Malta's current tourism offer.

Inclusivity

A similar argument based on the successful development of major events comes from Stevenson (1998) and Garcia (2011) who focus their observations based on 'grassroots consultations', which encourage inclusivity and benefits for local groups through tourism and inward investment. This would help offset the sometimes 'elitist' image of the RMSR event which, by and large, have tended to focus on private participation, upmarket events and 'by invitation only' venues. By organizing supplementary events in unison and making them accessible to the local public and tourists visiting would certainly provide additional benefits. One respondent also intimated that such events do not have to be solely maritime-related but could also be on shore such as dockside/waterfront festivals, thus giving the public the opportunity to get closer to enjoying the 'race spirit'. Such examples can be found already. The 'Taste of Tasmania' is a case in point. This event constitutes a vibrant food festival that takes place in Hobart which attracts volumes of tourists to the island and provides a culmination of the Rolex Sydney to Hobart race.

Seasonality and Over-Wintering

The concept of re-visitation, as explored for.example, by Robinson and Gammon (2004) in their conceptual analysis of primary and secondary motives for revisiting sports events was identified as a key advantage of the RMSR event. A couple of respondents stressed that they look forward to planning a holiday here with family as they feel Malta is particularly 'child-friendly' with most overseas visitors making a point of attending the event every year and some have done so for the past ten years and more.

This goes hand in hand with the development of the Blue Economy and associated sailing and yachting activities. In this respect Royal Malta Yacht Club (RMYC, 2013) has developed a calendar of yacht and sailing events which has been strengthened by the RMSR over the years. These now include:

May—The Bank of Valletta Round the Islands Race.
June—Slam Regatta and Citadel Ragusa Weekend.
July—Malta Syracuse Race.
August—Tommy Hilfiger Ramla Bay Regatta.
September—Oiltanking Vega Offshore Race.
November—IIG Gozo Weekend.
December—Medcomms Round Malta Race.

Together with such developments, the concept and opportunities for 'over-wintering' and extending the tourism season have been identified as key benefits. In this respect respondents stated that Malta as a destination is easily accessible, safe, offers a temperate climate all year-round, has adequate infrastructure, services and skilled labour. One local yacht executive interviewed also remarked on the uniqueness of such a combination and the additional advantages of a vibrant property market, an easy business setup environment and a favourable offshore tax regime. Thus, respondents suggested that over-wintering offers a broad range of opportunities. Such opportunities can in turn create new business, employment and new off-season markets if, as respondents suggest, the right building blocks are in place. In this respect, there were some negative responses related to poor quality basic infrastructure, limited stakeholder/community engagement and support which need to be addressed if such opportunities are to be fulfilled.

As Barker et al. (2002) suggest, in their research, which focused on the major impacts of hosting the America's Cup in the city of Auckland New Zealand, results can demonstrate significant economic benefits to the city and region. The results from respondents also infer such benefits. The RMSR can demonstrate that the broader promotion of Malta's maritime culture can assist with the development of ancillary events and activities giving added value and boosting local 'maritime' businesses and enterprise and in turn leading to general economic gain. These, for example, include additional interest in boat/yacht/sailing shows and superyacht conferences or exhibitions which facilitates interest in local expertise linked to shipbuilding, boat chandlery, refurbishment, maintenance and over-wintering facilities.

Other key advantages included broader investment opportunities, accessibility from air-connectivity of the islands from other European and even global destinations which provides a strong contributing factor towards considering Malta as a good wintering option. In this respect one respondent also made a point that the favourable climate throughout the winter months makes Malta an ideal weekend getaway from life in northern Europe. These are also sentiments that fit well with the main objectives of the 2012–2016 Tourism Policy for the Maltese Islands (Ministry for Tourism Culture and Education—MTCE, 2012) which goes someway to encourage and promote such activities and business development.

The Blue Economy, Building a Stronger Yachting Tourism Industry

The literature pertaining to sailing and yachting events demonstrates that the promotion and development of such activities and capabilities can contribute substantially towards building a stronger tourism product (EU, 2013, 2016). As such Malta has the capability to attract a number of other international sailing events and in turn host such events. Higham (1999) and Blain et al. (2005) also support such notions by suggesting that taking such an approach can capitalize upon existing infrastructure, lower investment costs, help minimize seasonality contributing to wider sustainability goals while in turn providing good marketing, branding and added financial benefits which is certainly the case for Malta.

In this respect respondents suggest that one of the prime benefits has been the worldwide media coverage that the RMSR has brought. This very much aligns to Jago et al.'s (2003) co-branding theory which suggests that pairing an event with a destination will engender some transfer of image between the event and destination brands thus establishing new potential market growth. Rolex's sponsorship of the RMSR and the benefits that can accrue for a destination such as Malta cannot be underestimated in this respect. As one of the yachting fraternities interviewed remarked, 'Developing the start into a top spectacle is of a great advantage to Malta as the Grand harbour setting remains so clear in every person's mind - the ultimate branding identity' (Fig. 11.2).

As many yachting executives commented, a race with a fleet of almost 90 yachts from 18 different countries obviously offers benefits from a tourism perspective. There are good opportunities to attract new market growth, particularly those associated with (i) yacht chartering and the Russian market or (ii) developing a broader event programme by initiating more inshore races and collaboration with other existing events. In the context of the RMSR, its contribution towards Malta's yachting tourism brand is regarded very highly as an ideal destination for such future development. Respondents intimated there have already been several enquiries from the 'Meetings, Incentives, Conventions and

Fig. 11.2 Grand Harbour setting: RMSR start October 2016 (*Source* The authors)

Events' (MICE) tourism business sectors with a view to organize associated conferences during race days. These are key advantages discussed by for example Allen et al. (2011) who stipulate and assess the very real socio-economic gains that hosting events can bring to a particular locality or destination. In this respect they illustrate that events have a 'power' to raise the profile of their host destinations, attract visitors, deliver economic benefits and create jobs.

These are notions and findings that have been explored and can be evidenced from the results of the RMSR research results. Figure 11.3 illustrates some of the key benefits that the RMSR research findings have revealed in this context.

Although not in any priority of order, the findings and points highlighted in Fig. 11.3 illustrate that economic benefits associated with destination branding, the creation of new markets, capitalizing on off-season opportunities, the development of MICE by-products/events and the creation of new maritime business opportunities have been real gains for enhancing Malta's tourism offer. This in turn has aided government

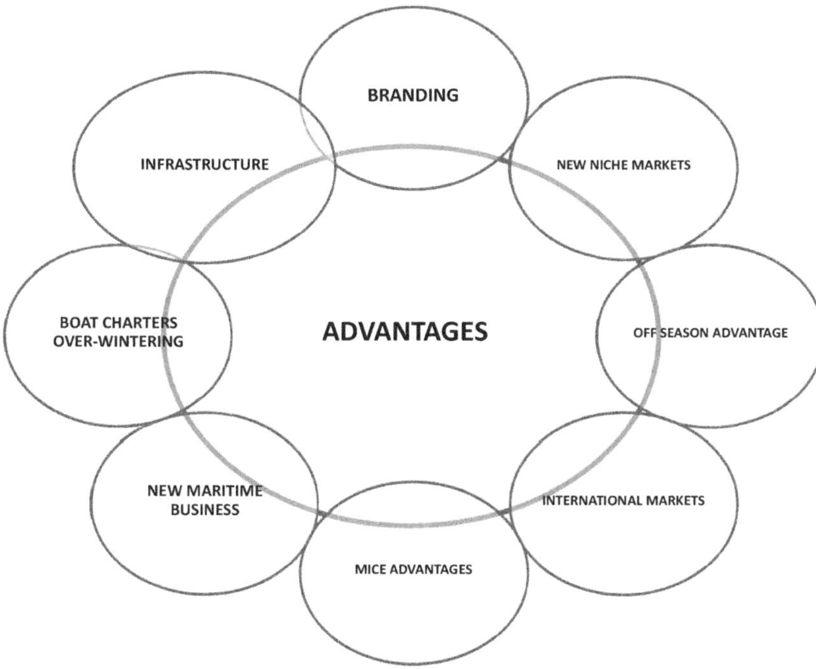

Fig. 11.3 Research findings—key advantages: THE RMSR EVENT (*Source* The authors)

policy and actions to redirect tourism from an over reliance on traditional mass markets to new opportunities for more diverse, upmarket and quality-based tourism which adds value to the existing market share. Events associated with yachting and sailing, such as the RMSR, thus demonstrate that at the luxury end of the market, specialist events focused on, for example, superyachts and luxury branding can offer a multitude of positive impacts for a host destination such as Malta.

Conclusion

Evidence from this research suggests that sailing and yachting has the capability of lending itself to various different areas of investment and business. The RMSR has contributed towards fulfilling tourism strategies promoted by the Malta Tourism Authority (MTA) for example, the '2012–2016 Tourism Policy' (2012) and the more recent launch of the Malta Tourism Authority 'Tourism Strategy 2021 - 2030 - Recover Rethink Revitalise - Malta' (MTA, 2021) both of which focus on diversifying Malta's existing tourism offer. In this respect, the RMSR event has gone some way to provide new incentives for broadening, diversifying and improving existing tourism operations and products. The RMSR event has provided Malta the chance to expand its tourism markets and assist in regenerating maritime traditions. In addition, the promotion of sailing and yachting has provided opportunity for further development of both the maritime and blue economies across the islands.

In summary, evidence suggests that the RMSR event has provided new perspectives to promote Malta's international tourism image. The RMSR, can clearly create opportunities for the establishment of new pioneering and inventive tourism thus assisting with reinvigorating traditional tourism markets such as Malta. In turn, this can also lead to the development of a stronger platform to promote wider sailing and yachting activities and maritime industries. Such developments can also have more wider implications, not only within a Maltese context, but also help stimulate wider interest in maritime traditions, maritime tourism and blue growth strategies across the Mediterranean region and internationally. In broader policy and management terms, the RMSR and its association with luxury branding, excellence and its emphasis on high quality, objectives, aspirations and outcomes has also certainly helped Malta steer a path to better quality tourism which gradually has led to refocus away from its reliance on traditional mass tourism.

As a consequence of the RMSR, new opportunities now present themselves to enhance Malta's tourism offer and provide a framework from which to build alternative tourism structures. These are currently changing and re-focusing attitudes on a better quality and the promotion of enhancement tourism infrastructures. These particularly now relate to the upgrading of marketing and destination branding focus, upgrading hotel accommodation, driving improvement for general hospitality, community and cultural experiences, improving tourist public realms and spaces and enhancing access, transport and mobility. The RMSR makes a small contribution to the overall tourism economy of Malta, nevertheless while benefits may still seem small from a larger perspective, such events and manifestations can provide a real catalyst from which new and unexpected tourism opportunities can develop and flourish.

References

Allen, J., O'Toole, W., Harris, R., & McDonnell, I. (2011). *Festival & special event management*. Wiley.

Arruda, N., & Ferrand, J. (2007). *The wind is calling—2007 ISAF Sailing World Championships*. Sete Mares.

Atwal, G., & Williams, A. (2009). Luxury brand marketing—The experience is everything! *Journal of Brand Management, 16*(5), 338–346.

Barker, M., Page, S. J., & Meyer, D. (2002). Evaluating the impact of the 2000 America's Cup on Auckland, New Zealand. *Event Management, 7*, 79–92.

Bickman, L. (1987). *Functions of program theory.* Retrieved January 22, 2021, from https://www.researchgate.net/publication/229635666

Bladen, C. (2021). The service semiotics of luxury events: An exploration for future research and events management industry practice. *Research in Hospitality Management, 11*(1), 53–57.

Blain, C., Stuart, E. L., & Ritchie, J. R. B. (2005). Destination branding: Insights and practices from destination management organizations. *Journal of Travel Research, 43*(4), 328–338.

Borg, T. (2012a). The middle sea race 1968–2011. *Spinnaker Magazine, 30*, 20–27.

Borg, T. (2012b, October 8). *Times of Malta. It is 90…and still counting*. Retrieved June 21, 2021, from https://timesofmalta.com/articles/view/It-s-90-and-still-counting.440144

EU. (2012). *Challenges and opportunities for coastal and maritime tourism in Europe.* Retrieved January 19, 2021, from http://ec.europa.eu/dgs/maritimeaffairs_fisheries/consultations/tourism/index_en.htm

EU. (2013). *Blue growth for coastal and maritime tourism in Europe.* Retrieved January 19, 2021, from https://ec.europa.eu/maritimeaffairs/content/european-maritime-day-2013-0_en

EU. (2014). *A European strategy for more growth and jobs in coastal and maritime tourism, EU/Europa, Brussels.* Retrieved March 9, 2021, from https://op.europa.eu/en/publication-detail/-/publication/88073fdf-4333-4d45-a012-aff4eef1bf3b

EU. (2016). *Study on specific challenges for a sustainable development of coastal and maritime tourism in Europe. EU/Europa, Brussels.* Retrieved March 9, 2021, from https://op.europa.eu/en/publication-detail/-/publication/ab0bfa73-9ad1-11e6-868c-01aa75ed71a1/language-en/format-PDF/source-search

Garcia, B. (2011). Urban regeneration, arts programming and major events. *International Journal of Cultural Policy, 10*(1), 103–118.

Getz, D. (2008). Event tourism: Definition, evolution and research. *Tourism Management, 29*, 403–428.

Getz, D., & Page, S. (2016). Progress and prospects for event tourism research. *Tourism Management, 52*, 593–631.

Global Yacht Racing. (2012). *Fastnet race*. Retrieved January 20, 2021, from https://globalyachtracing.com/events/fastnet-race/

Hall, C. (2001). Trends in ocean and coastal tourism: The end of the last frontier? *Ocean & Coastal Management, 44*, 601–618.

Higham, J. (1999). Sport as an avenue of tourism development: An analysis of the positive and negative impacts of sport tourism. *Current Issues in Tourism, 2*(1), 82–90.

Hughson, J. (2011). Sport in the "city of culture." *International Journal of Cultural Policy, 10*(3), 319–330.

Jago, L., Chalip, L., Brown, G., Mules, T., & Ali, S. (2003). Building events into destination branding: Insights from experts. *Event Management, 8*(1), 3–14.

Kane, M. J. (2010). Adventure as a cultural foundation: Sports and tourism in New Zealand. *Journal of Sport & Tourism, 15*(1), 27–44.

Macleod, D. (2006). Cultural commodification and tourism: A very special relationship. *Tourism Culture & Communication, 6*(2), 71–84.

Malta Independent. (2007, June 18). *Government plans to spend Lm300,000 on music festival and adverts on MTV.* Retrieved June 21, 2021, from https://www.independent.com.mt/articles/2007-06-18/local-news/Government-Plans-to-spend-Lm300,000-on-music-festival-and-adverts-on-MTV-175141

Malta Tourism Authority (MTA). (2012). *Malta tourism policy 2012–2016*. Ministry for Tourism, Malta.
Malta Tourism Authority (MTA). (2021). *Recover rethink revitalise—Malta tourism strategy 2021–2030*. Ministry for Tourism, Malta.
Ministry for Tourism Culture and Education (MTCE). (2012). *Tourism policy for the Maltese islands 2012/2016*. Retrieved January 22, 2021, from https://tourism.gov.mt/en/departments-sections-units/pages/departments-sections-units%20sub%20pages/eu%20affairs%20and%20policy%20development/tp12-16.pdf
New Zealand Tourism. (2021). *New Zealand—A sailing nation*. Retrieved March 10, 2021, from https://media.newzealand.com/en/story-ideas/new-zealand-a-sailing-nation/
Novelli, M. (2005). *Niche tourism: Contemporary issues, trends and cases*. Elsevier.
Phau, I., & Prendergast, G. (2001). Consuming luxury brands: The relevance of the "rarity principle." *Journal of Brand Management, 8*(2), 12–137.
Robinson, T., & Gammon, S. (2004). A question of primary and secondary motives: Revisiting and applying the sport tourism framework. *Journal of Sport & Tourism, 9*(3), 221–233.
Robinson, M., & Novelli, M. (2005). Niche tourism: An introduction. In M. Novelli (Ed.), *Niche tourism: Contemporary issues, trends and cases* (pp. 1–11). Elsevier.
Rolex. (2021). *Rolex race history*. Retrieved March 9, 2021, from https://www.rolexmiddlesearace.com/race/history
Royal Malta Yacht Club. (2012). *History*. Retrieved October 5, 2020, from https://www.rmyc.org/rmyc/history/
Royal Malta Yacht Club. (2013). Retrieved November 1, 2021, from https://www.rmyc.org
Stevenson, T. B. (1998). *International Review for the Sociology of Sport*. Retrieved January 22, 2021, from https://journals.sagepub.com/doi/abs/10.1177/101269098033003011
Sultana, W. (2009). *Times of Malta: The economic and branding benefits of the Rolex MSR*. Retrieved June 21, 2021, from https://timesofmalta.com/articles/view/the-economic-and-branding-benefits-of-the-rolex-msr.280812
Sultana, W. (2012). *Yachting in Malta* (10th ed.). Retrieved October 12, 2020, from https://issuu.com/anatlus/docs/yim2013-14
Trauer, B. (2006). Conceptualizing special interest tourism—Frameworks for analysis. *Tourism Management, 27*, 183–200.

Trauer, B., & Ryan, C. (2005). Destination image, romance and place experience—An application of intimacy theory in tourism. *Tourism Management, 26*, 481–491.

Vigneron, F., & Johnson, L. W. (2004). Measuring perceptions of brand luxury. *Journal of Brand Management, 11*(6), 484–506.

Visit Malta. (2021). *Recover, rethink, revitalise—Find out about the Malta Tourism Strategy, MTA*. Retrieved on March 9, 2021, from https://www.mta.com.mt/en/news-details/285

Weed, M., & Bull, C. (2009). *Sports tourism participants, policy and providers*. Elsevier.

Weiler, B., & Hall, C. M. (1992). *Special interest tourism*. Belhaven.

Part IV

Conclusion

12

Conclusion: Luxury Yachting—New Insights and Further Questions

Tatiana Gladkikh, Hugues Séraphin, Victoria Gladkikh, and Tan Vo-Thanh

This book is an attempt to shed more light on the luxury yachting industry which, despite its global presence and impressive net value, is not very well known to a wider society and not deeply explored and understood by researchers. The range of topics and chapters selected for

T. Gladkikh (✉)
The University of the Highlands and Islands, Perth, UK
e-mail: Tatiana.Gladkikh.Perth@uhi.ac.uk

H. Séraphin
Business School, University of Winchester, Winchester, UK
e-mail: hugues.seraphin@winchester.ac.uk

V. Gladkikh
Independent Luxury Yacht Practitioner, Dunkeld, UK

T. Vo-Thanh
La Rochelle Business School, Excelia Group, La Rochelle, France
e-mail: vothanht@excelia-group.com

T. Gladkikh et al. (eds.), *Luxury Yachting*,
https://doi.org/10.1007/978-3-030-86406-4_12

this book is united by the book's wide geographical span of the authors' experiences and locations, stretching from Australia and South-East Asia to the Mediterranean and the Americas. Our contributors have discussed luxury yachting from an international perspective and have brought their academic and practitioner insights gained from their research and professional work in various parts of the world.

Owing to the niche character of the industry which is not widely known and experienced by the majority of the population, the practitioner contributions are seen as of particular value to the reader as they provide an opportunity to glance 'below deck' and expose the operational side of the industry. At the same time, the academic analysis of the industry and its context brings additional benefits to the tourism research community as it enhances our knowledge of the issues related to luxury yachting.

The common thread observed across all the chapters is the recognition that the industry as Special Interest Tourism (SIT) has faced its own challenges and opportunities in the context of the global pandemic caused by the global spread and threat of COVID-19. While the pandemic generally has had a devastating effect on the tourism industry (Jamal & Budke, 2020), luxury yachting tourism appears to be a form of tourism that is still operating successfully, as remotely positioned yachts are seen as attractive isolated and contained islands offering a safe escape from the reality of the pandemic. Owing to the privacy and seclusion luxury yachting can offer, this type of Special Interest Tourism has enjoyed the demand for growth as more people seek safety in their tourist experiences. This raises the question whether this tendency for interest and demand for luxury yachting experiences could be contributing to overtourism, one of the most discussed issues in the current Tourism debate. To address this question, it is worth highlighting the fact that Special Interest Tourism has never been identified as a contributor of overtourism and has never been affected by overtourism-related criticism (Dodds & Butler, 2019; Milano et al., 2019). Arguably, SIT has been identified as a way of addressing the issue of overtourism (Séraphin & Yallop, 2021; Séraphin et al., 2020). Thus, the growth of the industry does not pose any threat towards the growing global phenomenon of overtourism.

Another common denominator among the contributions to this book is the issue of sustainability. It has been very loudly voiced by the practising professionals in particular, who are passionate about the sustainable future of the industry and who are searching for ways of addressing environmental and societal sustainability concerns in order to introduce relevant and innovative solutions into their daily operations. We have learnt from our contributors working in the industry that growing attention is being given now to sustainable practices of addressing the issues of pollution and sustainable food consumption. There are also noticeable developments in the field of employee rights and professional support through recognition of the challenging environment they work in. Further developments and actions, however, are needed to address these concerns more fully.

It is equally important to stress that luxury yachting tourism could be associated with sustainable form of tourism from a destination perspective. Favre (2017) argues that the main issue of destinations is that they do not know how to yield income out of the visitors they receive. The real problem lies in the fact that many destinations are still learning how to attract the right types of visitors, so that they would be committed to act as apostles for the destination and interested in bringing economic contribution even if they happened to be in that destination by chance (Favre, 2017). Luxury yachting could enable destinations to generate some significant income by developing a wider range of products, services and experiences aimed at guests and crew of luxury yachts, not only through supply of fuel and food but through building new sustainable facilities, developing wider networks of tourism providers, capitalizing on the ideas of geo-tourism, and organizing yachting events. As Séraphin (2020) and Vo-Thanh et al. (2021) argue, negative contexts, such as the COVID-19 pandemic, offer opportunities to review some exiting theories, practices and beliefs. Luxury yachting tourism addresses many of these points.

Another important consideration which this book has raised through its discussions is linked to the recognition of its global span of operations which take place in a truly transnational context. When the culture of the industry brings people from differing national, cultural and educational backgrounds together where they collectively face the same challenges while travelling globally, can we presume that luxury yachting carries the ability to act as a springboard for the development of a global mindset through the enhancement of shared values and morality leading to a united response to the global issues raised in the book, such as sustainability of our natural and social environment?

Can the multicultural character of the industry lead to the erosion of national identity (Gladkikh, 2018) and serve as an underlying fabric of a unique context of the new *global* culture and identity, where the mindset of those linked to the industry is aligned to that of the emerging transnational capitalist class (TCC) (Beck, 2006; Sklair, 2001). Could this context become a pioneering environment with universal values and morals which transcend national and cultural borders, and stereotypes? Can the industry become a vehicle for facilitating the emergence of a global society living 'in the global' (Freeland, 2007) with shared appreciation of the needs, challenges and values which we all share?

These questions demand further explorations which are outwith the scope of this book. However, by raising awareness of the industry, and by challenging and expanding our understanding of it, we hope to stimulate further enquiries into the world of luxury yachting and explore further its role in tourism, practice and context within which it operates.

References

Beck, U. (2006). *Cosmopolitan vision.* Polity Press.

Dodds, R., & Butler, R. (2019). *Overtourism.* De Gruyter.

Favre, C. (2017). The Small2Mighty tourism academy: Growing business to grow women as a transformative strategy for emerging destinations. *Worldwide Hospitality and Tourism Themes, 9*(5), 555–563.

Freeland, C. (2007, October 27/28). The no-nation generation. *Financial Times.*

Gladkikh, T. (2018). *The international business environment and national identity*. Routledge.

Jamal, T., & Budke, C. (2020). Tourism in a world with pandemics: Local-global responsibility and action. *Journal of Tourism Futures, 6*(2), 181–188.

Milano, C., Novelli, M., & Cheer, J. M. (2019). Overtourism and tourism-phobia: A journey through four decades of tourism development, planning and local concerns. *Tourism Planning & Development, 16*(4), 353–357.

Séraphin, H. (2020). COVID-19: An opportunity to review existing grounded theories in event studies. *Journal of Convention and Event Tourism, 22*(1), 3–35.

Séraphin, H., Gladkikh, T., & Vo-Thanh, T. (2020). *Overtourism: Causes, implications and solutions*. Palgrave Macmillan.

Séraphin, H., & Yallop, A. (2021). *Overtourism and tourism education*. Routledge.

Sklair, L. (2001). *The transnational capitalist class*. Blackwell.

Vo-Thanh, T., Vu, T.-V., Nguyen, N. P., Nguyen, D. V., Zaman, M., & Chi, H. (2021). How does hotel employees' satisfaction with the organization's COVID-19 responses affect job insecurity and job performance? *Journal of Sustainable Tourism, 29*(6), 907–925.

Index

T. Gladkikh et al. (eds.), *Luxury Yachting*,
https://doi.org/10.1007/978-3-030-86406-4

The manufacturer's authorised representative in the EU is Springer
Nature Customer Service Centre GmbH, Europaplatz 3, 69115 Heidelberg,
Germany. If you have any concerns regarding our products, please
contact ProductSafety@springernature.com

Printed and bound by CPI Group (UK) Ltd, Croydon, CR0 4YY
30/09/2026
02225499-0007